THE GALÁPAGOS

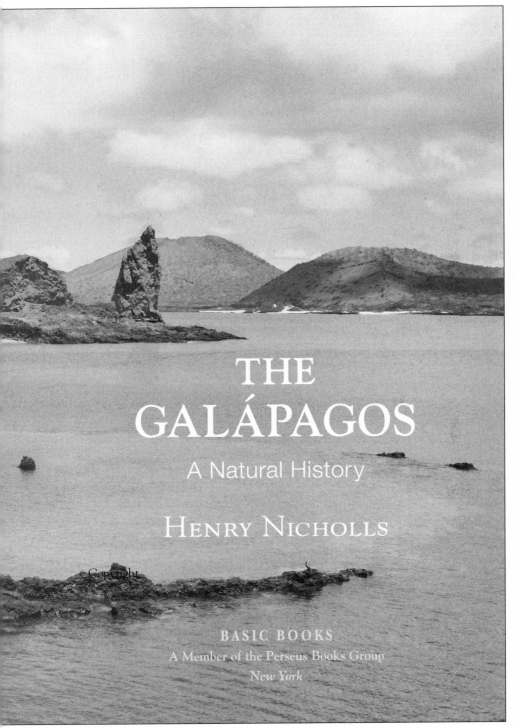

THE
GALÁPAGOS

A Natural History

HENRY NICHOLLS

BASIC BOOKS
A Member of the Perseus Books Group
New York

Copyright © 2014 by Henry Nicholls

Published by Basic Books,
A Member of the Perseus Books Group

All rights reserved. Printed in the United States of America. No part of this book may be reproduced in any manner whatsoever without written permission except in the case of brief quotations embodied in critical articles and reviews. For information, address Basic Books, 250 West 57th Street, 15th Floor, New York, NY 10107-1307.

Books published by Basic Books are available at special discounts for bulk purchases in the United States by corporations, institutions, and other organizations. For more information, please contact the Special Markets Department at the Perseus Books Group, 2300 Chestnut Street, Suite 200, Philadelphia, PA 19103, or call (800) 810-4145, ext. 5000, or e-mail special.markets@perseusbooks.com.

Book design by Trish Wilkinson and Cynthia Young
Typeset in Goudy Oldstyle

Library of Congress Cataloging-in-Publication Data
Nicholls, Henry, 1973–
    The Galapagos / Henry Nicholls.
        pages cm
    Includes bibliographical references and index.
    ISBN 978-0-465-03597-7 (hardcover)—ISBN 978-0-465-03595-3
    (e-book)   1. Natural history—Galapagos Islands.   2. Galapagos
    Islands—History.   3. Human ecology—Galapagos Islands—History.
    4. Galapagos Islands—Environmental conditions.   5. Environmental
    protection—Galapagos Islands.   I. Title.
QH198.G3N53 2014
508.866'5—dc23
                                                          2013049261

10   9   8   7   6   5   4   3   2   1

To the memory of Lonesome George,
for what he tortoise

The natural history of these islands is eminently curious, and well deserves attention..

— Charles Darwin, *Journal of Researches into the Natural History and Geology of the Countries Visited During the Voyage of H.M.S. Beagle,* 1845

Charles Darwin effected the greatest of all revolutions in human thought, greater than Einstein's or Freud's or even Newton's, by simultaneously establishing the fact and discovering the mechanism of organic evolution.

— Julian Huxley, "Charles Darwin: Galápagos and After," 1966

Only one English word adequately describes his transformation of the islands from worthless to priceless: magical.

— Kurt Vonnegut, *Galápagos: A Novel,* 1985.

No area on Earth of comparable size has inspired more fundamental changes in Man's perspective of himself and his environment than the Galápagos Islands.

— Robert Bowman, "Contributions to Science from the Galápagos," 1984

Contents

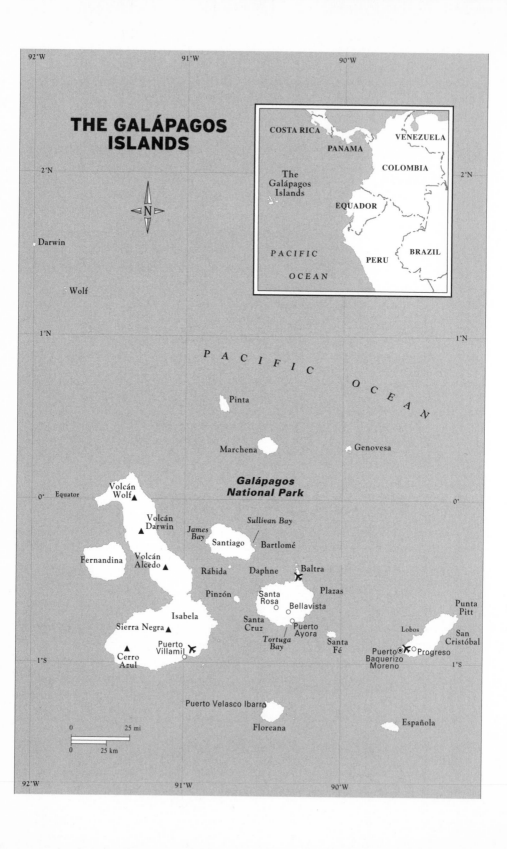

Prologue

On the morning of 7 December 1941, the Japanese bombed the United States naval base at Pearl Harbor in Hawaii, an event that brought the full force of the United States into World War II. It was also an event that played a major role in opening up the isolated Galápagos Islands to the rest of the world.

The Galápagos archipelago lies in the eastern Pacific Ocean, straddling the equator some 925 km off the west coast of South America. For the United States, this was the perfect location to establish a military base from which to defend the Panama Canal—a strategic lifeline—against a German or Japanese offensive. Reluctantly, Ecuador agreed to give the United States access to the Galápagos Islands for 'the establishment of such military bases as may be necessary'.

The Galápagos is made up of thirteen islands of notable size, ranging from the circular island of Genovesa (at just under 5 km in diameter) to the seahorse-shaped Isabela (more than 130 km from top to bottom). The other rocks, outcrops and islets in the archipelago—of which there are more than one hundred—would have been no place to build a military base. The US Navy Department's Office of Naval Intelligence was quick to produce a report on the Galápagos, identifying sources of freshwater (springs, lakes, streams, wells, pools), trails and roads, details of human settlements and their inhabitants, possible anchorage sites for vessels, seaplanes and submarines, and the most suitable spots to situate an airfield. The classified report identified two possible sites for a landing strip on Baltra, one of the smaller islands located in the centre of the archipelago, adding that 'both require considerable clearing'.

By April 1942, crushed lava had been compacted and sealed with hot asphalt and the first airstrip was ready to receive its first plane. The US forces began to arrive a month later and the runway remained in continual use until Ecuador ushered the North Americans out of their territory after the war. Still, it's an important moment in the strange history of these islands, one that paved the way—quite literally—for the arrival of commercial flights to the Galápagos (although visitors flying into Baltra today have a smoother landing on a second runway constructed on the other side of the island).

Before the United States gave up its stronghold, however, Franklin D. Roosevelt expressed his vision for the future of the archipelago. From a memorandum written to Secretary of State Cordell Hull in February 1944, it's clear that the Galápagos mattered to the president. 'These Islands represent the oldest form of animal life and should, therefore, be preserved for all time as a kind of international park,' he wrote. 'I have been at this for six or seven years and I would die happy if the State Department could accomplish something on it!' he urged a little later.

FIGURE I. *Franklin D. Roosevelt.* The US president looks chuffed with his catch off Santiago in July 1938. *US Naval Historical Center.*

Unfortunately for Roosevelt, who died a year later, just before the end of hostilities, he never got to realise this dream. In time, though, Ecuador embraced the idea of protecting these islands and today the Galápagos archipelago matters to us all. It matters to those who live there (the plants, the animals, the humans whose lives depend on it). It matters to Ecuador (whose tourism industry is based on it). It matters to hundreds of thousands of non-Ecuadorians (who have had the good fortune to visit). It matters to the rest of the world as a model system (for what it might yet teach us).

Let me expand.

The landscape is both hostile and beautiful; the wildlife is sparse yet striking. Scientists have documented just over 4,000 species native to the Galápagos, around 40 percent of them endemic, found here and only here. It's not just this 40 percent, these 1,600 species, that owe their existence to the Galápagos. There are now around 30,000 people living in this far-flung constellation of islands whose livelihoods depend—either directly or indirectly—on the integrity of the ecosystem and the tourism-based economy it fuels.

More widely, Ecuador cares deeply about this otherworldly territory it acquired in 1832. For its small size (occupying less than 2 percent of South America), Ecuador has an impressive range of habitats, making it one of the most biodiverse countries in the world (thought, for instance, to be home to almost half of all the bird species on the continent). Yet when it comes to international tourism, these other offerings just can't compete with the pulling power of the Galápagos. For many travellers, it's the sole reason for paying Ecuador a visit at all.

For those who have been lucky enough to visit, a trip to the Galápagos is likely to be up there amongst the most memorable experiences of their lifetime. The animals show no prejudice, no fear, but accept humans for what they are, just another species attempting to live in this inhospitable outpost. Experiencing this equanimity with nature is so moving that it has the power to alter the course of human lives, to transform the way we think about our place in the world and the way we behave towards its other inhabitants, human and non-human alike. Since tourism to the islands began less than fifty years ago, over 1.5 million people have had the chance to see these 'Enchanted Islands'. In 2003, I became one of them, a visit that inspired my first book in which I used the tale of the Galápagos' most famous resident (a solitary giant tortoise called Lonesome George) to explore the challenges facing conservationists in the archipelago and beyond. I have been back again, but my main contact with the islands has been from afar: I have continued to write about the Galápagos; I became an ambassador for the Galápagos Conservation Trust and the editor of its magazine, *Galápagos Matters*. Hardly a day goes by when I do not think about these wonderful islands.

If you haven't been fortunate enough to visit, the Galápagos still matters. There are several reasons. By virtue of the ecological erosion that humans have caused elsewhere, the Galápagos now stands out. Of those 4,000 native species, only 17 are known to have gone extinct. This makes the Galápagos the most pristine archipelago to be found anywhere in the tropics. It is so remote, so relatively untouched, that the act of wading ashore to one of its islands can feel like you are the first to do so. In an increasingly disturbed world, it will be of increasing value that we still have places like the Galápagos that offer this kind of experience. There just aren't many of these left.

There is another, farther-reaching reason why we should value the Galápagos, even if only from a distance. Its relatively simple and pristine nature makes it a brilliant place to get to grips with the relationship between different species, uncomplicated by the heavy hand of humankind. In particular, the isolation of the Galápagos from the South American continent and the proximity of the islands to one another make it a perfect place to detect the origin of new species. The Galápagos remains one of the best places to study the process of evolution in the field, how natural selection can result in biological novelty.

The contribution the Galápagos made to the genesis of Charles Darwin's ideas on evolution is another reason to pay these islands special attention. The naturalists he inspired, a succession of Darwin anniversaries and our eagerness for a simple story mean that the Galápagos has become intertwined with Darwinian evolution. Learning about the Galápagos then is a great way to learn about what is arguably the most influential idea in the history of human thought. The Galápagos finches have been flitting around the schoolrooms of the world for decades, avian evangelists for the power of evolution by natural selection.

Just as the Galápagos became a model for Darwin, so it is a model for anyone who cares about our future. In spite of its relatively pristine state, humans have had a profound impact on these islands, eating its tortoises in the hundreds of thousands, hacking down the highland habitats to make way for farms and introducing non-native species with devastating consequences. On the plus side, there have been plenty of achievements to celebrate over the last century, notably the designation of 97 percent of its landmass as national park in 1959, the creation of the Galápagos Marine Reserve in 1998, and some of the world's most ambitious attempts at ecological restoration.

In recent years, there has been increasing talk of the Galápagos playing a kind of inspirational, transformative role, serving as a perfect testing ground for what is known as sustainable development. As yet, there is only a handful of cases that suggest humans are capable of development without destruction, and these are relatively small-scale. The Galápagos offers a crucial test of a truly important idea. In an archipelago stamped with UNESCO's World Heritage seal of approval, blessed with the Darwin brand, bolstered by solid international support and fed by a steady stream of tourist dollars, is it really beyond the wit of humankind to meet the wants of the current generation without compromising the needs of the next? The Galápagos really matters because what happens in these islands is an honest, unfalsifiable look at the future that faces our children, our grandchildren and our species.

* * *

I have structured my account of the Galápagos around the natural history for which the islands are famous. So we begin with volcanoes, for without them there would be no islands at all. The bleak, lava-strewn world is certainly impressive. The bishop of Panama, one of the first people to set foot in the Galápagos in 1535, felt as though God had showered stones upon the landscape. What soil there is, he wrote, 'is like dross, worthless, because it has not the power of raising a little grass, but only some thistles.' Exactly three hundred years later, Darwin found the vista similarly hostile. But unperturbed, he began to chip away at exposed strata and collect samples of volcanic rock. His ideas anticipated the notions of continental drift, plate tectonics and deep-seated, superheated hotspots that account for the peculiar formation we call the Galápagos.

We then consider the marine realm that thrives in the nutrient-rich waters around the islands. The American naturalist and explorer William Beebe became the first person to dive in Galápagos waters almost a century ago, and it is through the reinforced glass windows of his bulky helmet that we get a first glimpse of the thrills that lie beneath the waves. It was also near the Galápagos that scientists first photographed rifts in the ocean floor and found signs—extraordinary signs—of life several kilometres below the surface. The Galápagos Marine Reserve ring-fences all these riches. There are fish, of course, including spectacular, vast schools of hammerhead sharks. But it's also home to sea turtles, sea lions, fur seals, dolphins and whales.

With the power of flight, seabirds were quick to colonise, using the bare rocks as a nesting base and the ocean as a larder. There are comical boobies, awkward pelicans, magnificent frigatebirds, huge albatrosses, sleek petrels and the now flightless cormorants. We discover why the blue-footed booby's feet are so coloured, why the male frigatebird should pump up his neck like a balloon and why on earth cormorants should have given up on a trait as wonderful as flight.

Vegetation, meanwhile, was taking root: first lichens, then hardy weeds, then bigger plants and trees, species that in concert with the way that rain falls at different altitudes mark out three principal habitats in the Galápagos. There are the mangroves that dominate the coastal zone, their gnarled roots sucking moisture from the salty ocean and their bright-green leaves throwing up a magical boundary between sea and land. There is the arid, scrubby hinterland that Darwin felt contained 'wretched-looking little weeds' but also the marvellous sentinel-like *Opuntia* cacti. Further up, on the damp, cloud-covered slopes, there is lush and humid rainforest with stands of native *Scalesia* trees that have been on a truly remarkable evolutionary journey.

Invertebrates began to settle in, beetles, weevils and snails all diverging into myriad different species on different islands. There are butterflies and moths too, ants and a bee, spiders and ticks, all less conspicuous than the more famous giant tortoises but a crucial part of the Galápagos story nonetheless.

Although not particularly spectacular to look at, the Galápagos finches are rightly the best-known land birds in the archipelago. Research conducted over the last thirty years has turned these species into one of the finest examples of evolution by natural selection. More recently, the Galápagos mockingbirds have been receiving some rightful attention for the role they played in switching Darwin on to the notion that species can change—evolve—over time. These emblematic species aside, plenty of other land-based birds have made the Galápagos home, including flamingos and rails, doves and hawks, cuckoos and warblers.

If it's reptiles that are of most interest, then flip straight to this chapter. The marine iguanas that clothe every shoreline in the Galápagos have the most immediate impact on the casual visitor. Darwin was fascinated by these 'imps of darkness' and their ability to swim, to dive and to feed off the algae on submerged rocks. Recent research has shown them to be more impressive still, with individual animals able to change their body size, shrinking their skeleton when food is scarce and growing it again in times of abundance. There are land iguanas, lizards, geckos and snakes, but it's the gargantuan tortoises that are the most famous creatures in the Galápagos, the reptiles that give the islands their collective name. Genetic studies have helped to map out the origins of the different tortoise lineages, of which four are known to have gone extinct. More recently, behavioural work is getting to grips with their patterns of migration and the significant role they play in the dispersal and fertilization of seeds.

With the wonder and importance of these rocks, plants and creatures properly explained, it will be time to turn to more recent arrivals: *Homo sapiens* and other non-native species. In the centuries that followed the first human footfall in the islands almost five hundred years ago, pirates came, then whalers, then scientists and settlers. Darwin's visit in 1835 proved to be a pivotal moment for how humankind has come to view these islands. Inspired by tributes to Darwin in 1959, marking 150 years since his birth and the one hundredth anniversary of the publication of *On the Origin of Species*, Ecuador declared the Galápagos as its first national park. This step, which set the stage for tourism to the islands, has clearly brought a huge number of challenges, not least an established population of over 30,000, visitors approaching 200,000 in number every year, the influx of non-native species that threaten to unpick the ecological fabric of the Galápagos and a now ever-present miasma of inefficiency and corruption.

*   *   *

Charles Darwin will be a constant companion throughout. For those who do not buy into his views on the origin of species, this need not be cause for concern. The reason for paying special attention to Darwin is that his ideas have had a major impact on the way that humans see the natural world in general and the Galápagos in particular. Before HMS *Beagle* sailed into Galápagos waters in 1835, most visitors considered these islands to be some kind of godforsaken hell on earth. In its wake, perceptions began to change, and today the archipelago is widely perceived to be the closest thing we have to paradise. Darwin is crucial to that transformation from hell to paradise, absolutely fundamental to the modern identity of the islands. If we ignore the significance the Galápagos held for Darwin, we will have failed to appreciate the true importance of these islands.

In what follows, I will attempt to capture what is at stake: the awe-inspiring landscapes, the understated flora, the stunning, sometimes freakish wildlife and, crucially, the origin of new species. I will outline the immense challenges facing the islands and consider the future that lies ahead. I will not be writing about all the things to be seen in the Galápagos. That would be boring. But in cherry-picking, I hope I will have written an entertaining and enlightening account that distils the spirit of this remarkable place.

FIGURE 2. *Charles Darwin as a youthful explorer.* This portrait by English artist Thomas Herbert Maguire was drawn a decade after Darwin's return from the Galápagos. *US National Library of Medicine.*

Editorial Note

I refer to the Galápagos in the singular (as in 'the Galápagos Archipelago is very remarkable'). I am applying the accent to the word 'Galápagos' throughout (and to islands like Española, San Cristóbal, Pinzón, and Santa Fé and phenomena like El Niño and *garúa*); for the sake of consistency, I have also done so in quotations and references where the accent may have been dropped. Over the course of the last several hundred years, many of the islands have been named and renamed, some being christened more than ten times; in order to minimise confusion, I am using the modern Ecuadorian taxonomy except where absolutely necessary.

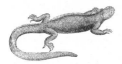

Chapter 1. Rocks

For the US forces stationed on Baltra during World War II, the Galápagos was about as far from paradise as they could imagine. They called it 'The Rock', a wry nod to the infamous prison Alcatraz in San Francisco Bay. When First Lady Eleanor Roosevelt reflected on her morale-boosting visit to the Galápagos in 1944, she acknowledged the backbreaking slog required to establish a military base in such an inhospitable terrain. 'It is as though the earth had spewed forth rocks of every size and shape and, as one man said, "You remove one rock, only to find two more underneath."' For the troops, 'it must be one of the most discouraging spots in the world', she wrote. At the same time, however, Mrs Roosevelt could see how this same landscape might grab the attention of those of a scientific bent. 'To a geologist, I'm sure it would furnish several years of absorbing work.'

The most celebrated visitor to the Galápagos, Charles Darwin, spent only five weeks in the islands, but he'd have gladly spent more. In 1835,

Darwin was far more interested in rocks than in living creatures and still believed in the creative powers of God. 'Geology is a capital science to begin, as it requires nothing but a little reading, thinking & hammering,' he wrote from his cramped cabin on board HMS *Beagle* as it lay anchored in the harbour at Lima, Peru. Of what remained of the voyage—across the Pacific and then home to England—it was the prospect of studying the Galápagos rocks that excited him most. 'I look forward with joy & interest to this . . . for the sake of having a good look at an active Volcano,' he wrote. 'Although we have seen Lava in abundance, I have never yet beheld the Crater.'

All this pent-up excitement erupted on the afternoon of 16 September 1835, when HMS *Beagle* anchored off the north-western tip of San Cristóbal in the east of the archipelago and Darwin had a chance to venture ashore. Although he was clearly thrilled to get his geologist's hammer out and start tapping away at the rocks, the island's landscape as a whole was foreboding. 'A broken field of black basaltic lava, thrown into the most rugged waves, and crossed by great fissures, is everywhere covered by stunted, sunburnt brushwood, which shows little signs of life,' he wrote in his *Journal of Researches*. In a rather wonderful passage, he went on to liken the 'strange Cyclopean scene' to his native Britain then in the throes of the Industrial Revolution. 'From the regular form of the many craters, they gave to the country an artificial appearance, which vividly reminded me of those parts of Staffordshire where the great iron-foundries are most numerous.'

The American author Herman Melville, who passed through the Galápagos on board the whaler *Acushnet* in the early 1840s (a trip that

---

FIGURE 1.1. **Right, *four views of the Galápagos*.** Philip Gidley King (a midshipman on the *Beagle* with whom Charles Darwin shared a cramped cabin) made many snapshot-like sketches during the voyage. These drawings, included as engravings in Robert FitzRoy's *Narrative*, show Floreana (Charles Island), the approach to San Cristóbal (Chatham Island), a close-up of Freshwater Bay on San Cristóbal, and Isabela (Albemarle Island). *Reproduced from* Robert FitzRoy, Narrative of the Surveying Voyages of His Majesty's Ships Adventure and Beagle . . . (London: Henry Colburn, 1839).

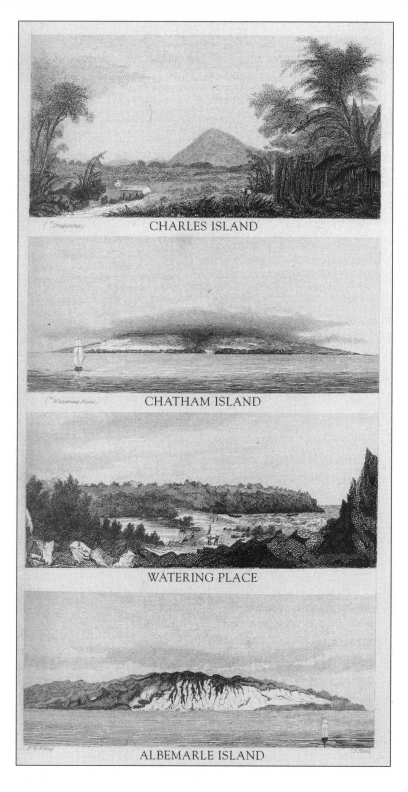

CHARLES ISLAND

CHATHAM ISLAND

WATERING PLACE

ALBEMARLE ISLAND

he would draw on for his most famous novel, *Moby-Dick*), drew upon and embellished Darwin's account. In *The Encantadas, or Enchanted Isles*, a series of ten literary sketches of the islands he published in 1854, he wrote, 'In many places the coast is rock-bound, or, more properly, clinker-bound; tumbled masses of blackish or greenish stuff like the dross of an iron furnace, forming dark clefts and caves here and there, into which a ceaseless sea pours a fury of foam, overhanging them with a swirl of gray, haggard mist, amidst which sail screaming flights of unearthly birds heightening the dismal din.'

First Impressions

Rather interestingly, the very first humans to step on Galápagos rocks had a similar reaction. On 23 February 1535, a galleon set sail from Panama in Central America and headed south. Its mission on behalf of His Sacred Imperial Catholic Majesty King Charles I of Spain was to convey Spain's head honcho in South America, Bishop of Panama Tomás de Berlanga, to Peru (to check up on the state of play following conquistador Francisco Pizarro's romp across the continent and the execution of the Incan emperor Atahualpa a few years earlier). The voyage should have taken a couple of weeks at most, but on reaching the equator, the galleon experienced 'a six-day calm'. Sucked out into the Pacific Ocean on the Panama and Humboldt Currents that merge at the equator, de Berlanga found himself in a cartographic void.

After several days helplessly drifting and with supplies of freshwater running perilously low ('there was enough water for only two more days'), it looked as though God had forsaken the bishop and his men. But on 10 March, an island—most likely Española in the south-east of the Galápagos archipelago—came into view. Lowering the lifeboat, a party went to search for water and to find grass for the horses on board. They came back disappointed.

Another larger and more mountainous island—probably Floreana—looked more promising, and 'thinking that, on account of its size and monstrous shape, there could not fail to be rivers and fruits', de Berlanga set a course for it. When they finally anchored, everyone—including the bishop—staggered ashore desperate for water. Some men set about trying to dig a well, but 'there came out water saltier than that of the sea'.

Others who went inland in search of a spring or stream 'were not even able to find even a drop of water for two days'. The reluctant explorers became so parched they resorted to tearing apart the cactuses abundant on the lower slopes of most of the Galápagos volcanoes. 'Although not very tasty, we began to eat of them, and squeeze them to draw all the water from them, and drawn, it looked like slops of lye, and they drank it as if it were rose water.'

Being a good Catholic, the bishop insisted on holding mass to mark Palm Sunday. As far as he was concerned, this act of faith—and the prayers he presumably offered up to his deity—resulted in a group of his men stumbling on freshwater soon afterwards. 'The Lord deigned that they should find in a ravine among the rocks as much as a hogshead of water, and after they had drawn that, they found more and more.'

Although this island had saved his life, de Berlanga did not give it the most glowing write-up: 'On the whole island I do not think that there is a place where one might sow a bushel of corn, because most of it is full of very big stones, so much so that it seems as though at some time God had showered stones.' What earth there is, 'is like slag, worthless,' he wrote, anticipating the similarity that Darwin would draw with Britain's industrialising heartland exactly three hundred years later.

Crack of Doom

The Galápagos landscape owes its striking appearance to the archipelago's volcanic origins. Darwin's expectation of seeing an active volcano in the Galápagos was probably raised by exciting accounts of the youngest island—Fernandina—blowing its top a decade earlier.

The American explorer Benjamin Morrell could not have been better placed to witness this dramatic event, sailing his vessel, the *Tartar*, into Isabela's Banks Bay on 10 February 1825. As luck would have it, he also had a rather nice way with words. In the middle of the night a few days later, 'while the stillness of death reigned everywhere around us, our ears were suddenly assailed by a sound that could only be equalled by ten thousand thunders bursting from the air at once; while, at the same instant, the whole hemisphere was lighted up with a horrid glare that might have appalled the stoutest heart!' he wrote. Less than 20 km away, Fernandina had suddenly 'broken forth with accumulated vengeance'.

This 'crack of doom' soon brought everyone on deck, where the men 'stood gazing like "sheeted spectres," speechless and bewildered with astonishment and dismay. The heavens appeared to be in one blaze of fire, intermingled with millions of falling stars and meteors; while the flames shot upward from the peak . . . to the height of at least two thousand feet in the air.'

By about four in the morning, 'the boiling contents of the tremendous caldron had swollen to the brim, and poured over the edge of the crater in a cataract of liquid fire. A river of melted lava was now seen rushing down the side of the mountain, pursuing a serpentine course to the sea,' wrote Morrell. The 'dazzling stream', he judged, was about a quarter of a mile wide, 'presenting the appearance of a tremendous torrent of melted iron running from the furnace.'. The 'flaming river' broke its banks in several places, sending fiery branches in every direction across the landscape, 'each rushing downward as if eager to cool its temperament in the deep caverns of the neighbouring ocean'. And when the lava met the water, the uproar was 'dreadful indeed'. 'The demon of fire seemed rushing to the embraces of Neptune; . . . The ocean boiled, and roared, and bellowed, as if a civil war had broken out in the Tartarean gulf.'

Morrell had the presence of mind to collect some data, recording the temperature of the sea and air as the drama unfolded. His baseline measurement, taken an hour after the first explosion and before the lava had begun to spill over the rim of the volcano, was fairly typical for the time of year with the water at around 16°C and the night air a balmy 21°C. Some six hours later, with the eruption 'still continuing with unabated fury', the temperature of the water had rocketed to an incredible 37°C and the air was now an oppressive 45°C. This was clearly alarming because Morrell and his men were trapped. 'Not a breath of air was stirring to fill a sail, had we attempted to escape,' he wrote. By the time the atmosphere had reached an unbearable 50°C, the glue-like resin holding the vessel together had started to run, tar was dripping from the rigging and Morrell and his men were struggling to breathe. Stripping off and jumping into the water would have offered them no respite. At over 40°C, it would have been like diving into a scorching bath.

Thankfully, 'a breath of a light zephyr from the continent, scarcely perceptible to the cheek' began to strengthen, and at last Morrell was able to weigh anchor. The wind created a new problem, though, spreading what seemed to be 'a mass of flame' north of Fernandina, barring a safe passage into the open Pacific to the west. The only option was to head south in an effort to get upwind of the volcano, and this meant passing within a few kilometres of Fernandina's molten shoreline.

At the narrowest point, Morrell found the water to be marginally hotter than the air, 'almost boiling' at over 60°C. 'I became apprehensive that I should lose some of my men, as the influence of the heat was so great that several of them were incapable of standing,' he wrote. The temperatures did ease once they were south of Fernandina, but they pushed on to the safety of Floreana. From there, some 80 km away, Fernandina's crater still appeared 'like a colossal beacon-light, shooting its vengeful flames high into the gloomy atmosphere, with a rumbling noise like distant thunder.'

Although Melville didn't witness anything like the eruption experienced by Morrell and his men, Fernandina clearly got his imagination going. 'There is dire mischief going on in that upper dark,' he wrote in the fourth of his ten sketches. 'There toil the demons of fire, who, at intervals, irradiate the nights with a strange spectral illumination for miles and miles around.'

Given Darwin's eagerness to see an active volcano in the Galápagos, he probably left a little deflated. The best he managed was a fumarole on neighbouring Isabela, where 'high up, we saw a small jet of steam issuing from a Crater'. Hardly earth-shattering. But this didn't stop him taking the first serious look at the geology of these islands. Everything he saw strongly suggested that the landscape was not as static as it might seem at first glance.

Faulting

In South America, Darwin had hit upon the idea that whole chunks of rock had to have been elevated at some time in the past. At his first stop-off in the Galápagos on San Cristóbal, he found an exposed cliff face that contained sandstone and the fossilized remains of limpets, mussels and molluscs. The fact that it was now a couple of feet above the high-tide mark was, he considered, 'proof of elevation to a small degree within recent times.' Of course, sea level might have dropped, leaving the marine deposits high and dry, but he thought this less likely.

The most striking example of uplift is at Urbina Bay on the west coast of Isabela. In 1954, a fisherman noticed a white stretch along the shoreline that had not been there before and, upon closer scrutiny, found an eerie landscape strewn with decomposing creatures and an unbearable stench. It's thought that a single volcanic event caused the ocean floor to rise—almost instantaneously—by around 5 m, exposing some 6 km of reef and stranding sea creatures in the process. There is further evidence of uplift at nearby Punta Espinoza on Fernandina (a projection of lava most likely created in the 1825 eruption witnessed by Morrell), where the tourist landing dock now stands completely out of the water at low tide as a result of a volcanic event in the 1970s.

Such movement of large chunks of rock is made possible by weaknesses, or 'faults', in the earth's crust. Much of the north-east of Santa Cruz appears to have experienced a bit of uplift. In fact, the incredibly flat surface of Baltra (one of the reasons it's such a good site for an airport runway) suggests it rose up from beneath the sea. The presence of shallow-water marine fossils embedded in the layers of exposed lava on North Seymour and Las Plazas indicates these islets are the results of uplift too.

If there's uplift, it stands to reason that blocks of crust might also sink. This is harder to demonstrate, but the main town of Puerto Ayora certainly looks as though it's nestled in a hollow caused by just such subsidence. At the southern edge of town, the footpath towards Tortuga Bay zigzags up a nearly sheer escarpment around 20m high that runs from the north-west to the south-east. From the top, it's possible to make out a parallel fault line on the far side of town. The land in between seems to have slipped, creating the perfect natural depression for all those houses.

The Rocks

Darwin also paid close attention to the makeup of the rocks, from which he began to infer much about the action of volcanoes. Whilst on Santiago, for instance, he noticed that the rocks on the lower slopes were darker and denser than those at the verdant summit. His explanation was simple. The only way a volcano would throw out lava of different densities, he reasoned, was if it had a large internal chamber of molten rock. This would allow lavas of different composition to emerge in one eruption, the densest near the bottom of the chamber bursting out around the flank of the volcano and the least dense oozing from the top (see Appendix C, Figure 1).

In the course of an eruption, the evacuation of this internal chamber can result in the formation of a caldera, the bowl-shaped depression in the centre of a volcano. The eruption of Fernandina in 1968 caused its magma chamber to collapse and the floor of its caldera to fall more than 200m. It's not possible to peer into this steaming basin, but the Galápagos National Park does have a tourist trail to the rim of the even larger caldera of Sierra Negra on Isabela.

The emergence of lava from around the periphery of the main vent results in a stunning range of other structures. For the sake of convenience, geologists like to bundle these up into different categories, though in reality small variations in chemical composition, temperature, pressure, eruption rate and the presence of water contribute to a nearly infinite diversity of rocky forms, each sliding imperceptibly into the next. Spatter cones are pretty much as they sound, soft and sticky lumps of lava that fly up into the air and return to earth with a splat, eventually building to leave an irregular cone-shaped blot on the landscape. Cinder cones are similarly explosive but with more gas involved, resulting in solid chunks of pumice-like rock raining down into a reasonably orderly cone. Tuff cones are more regular still, made from fine grains of volcanic ash. Seawater is a crucial ingredient in their formation, another of Darwin's sleuth-like deductions. If water washes into an open vent, it is transformed to a superhot steam that shoots upwards, flinging minute grains of lava into the air. As anyone who has let sand sift into a pile on the ground knows, these grains pile up into a perfect cone.

Whilst on San Cristóbal, Darwin marvelled at this tormented landscape.

> One night I slept on shore on a part of the island, where black truncated cones were extraordinarily numerous: from one small eminence I counted sixty of them, all surmounted by craters more or less perfect. . . .

The entire surface of this part of the island seems to have been permeated, like a sieve, by the subterranean vapours: here and there the lava, whilst soft, has been blown into great bubbles; and in other parts, the tops of caverns similarly formed have fallen in, leaving circular pits with steep sides.

Though the particular 'craterized district' that Darwin was describing here is out of bounds to tourists, there are lots of other sites where similar structures are visible. There are few better than those glimpsed from the top of Bartolomé, a spot that looks out on one of the most famous and photographed panoramas in the Galápagos. Gazing east, there are the pock-marked remains of several spatter cones. To the west, there are remains of an eroded tuff cone known as Pinnacle Rock. Over the hummock of tuff at the other end of Bartolomé, looking across to the far bigger island of Santiago, there's a prime example of a cinder cone in the middle distance.

When the lava rolls in a controlled, textbook style towards the sea, it can take on one of two main forms. When there's a lot of lava and it flows rapidly down the slope, it tends to forms fields of black rubble known as 'a'ā (a Hawaiian word meaning 'hurt' and one of the most useful Scrabble words you'll come across). When the Galápagos National Park Service (GNPS) began to select its visitor sites, it sensibly avoided fields of 'a'ā, but it's pretty much everywhere (as on the steep eastern slopes of Isabela's Wolf Volcano, for instance).

When the lava is flowing more slowly, it's more likely to result in rippled sculptures of extraordinary beauty known as pahoehoe (another Hawaiian term, meaning 'smooth'). One of the best places to see pahoehoe is at Sullivan Bay on Santiago, just across the water from Bartolomé. Darwin saw it on the other side of the island at James Bay. There, between the visitor sites of Espumilla Beach and Puerto Egas, is a vast field of pahoehoe lava, which in places Darwin felt resembled 'folds of drapery, cables, and . . . the bark of trees'. It's also in this flow of lava that the Norwegian explorer Thor Heyerdahl found fragments of marmalade jars in 1953. These, he reckoned, had been stashed there by buccaneers in 1683, suggesting the eruption that caused this lava field occurred some years later. This, quipped one witty geologist, is 'one of the rare uses of marmalade pots in volcano-chronology.'

It's pahoehoe lava that's also responsible for leaving tunnels through the landscape. As a thick stream rolls down the slope, the outer layer in contact with the air can form a skin whilst the molten rock continues to run inside. When the flow ceases, it leaves a long and cavernous tube. In the Santa Cruz highlands, there are several farms where it's possible to walk the length of such lava tubes. One of them runs directly beneath the road between Puerto Ayora and Bellavista.

Weathering

Whilst volcanic activity and faulting can cause nearly instantaneous alterations of the Galápagos landscape, the lapping of waves, the beating of the wind and the occasional deluge result in change on a much longer timescale. In spite of the uplift Darwin had documented on San Cristóbal, he was aware the islands were weathering away. In particular, he noticed the effects that the elements were having on the sandy tuff cones. Of the twenty-eight such cones he took a look at, all of them—without exception—'had their southern sides either much lower than the other sides, or quite broken down and removed.' Given the way the elements consistently batter the southern coasts of all islands, 'this singular uniformity in the broken state of the craters, composed of the soft and yielding tuff, is easily explained,' he wrote.

The most striking example of this weathering—certainly one that impressed Darwin—is at Tagus Cove on the west coast of Isabela. Indeed, it's only possible to sail into this popular visitor site because the southern rim of this cone has weathered away to let the sea flow into its centre. When pirates moored their vessels at this spot, their anchors would have been resting on a spot through which lava spewed in the long-distant past. All around the walls of the cove, layer upon layer of lava expose its repetitive and violent volcanic origins.

El León Dormido, or The Sleeping Lion, a popular dive site just off the north shore of San Cristóbal, offers another illustration of weathering. Darwin guessed that the amorphous mass of rock once filled the central hollow of a cone, the sloping walls of which have long since worn away.

FIGURE 1.2. ***Tagus Cove on Isabela.*** Darwin figured that this sheltered cove had to be the result of weathering, with the elements battering down the southernmost side of what had once been an intact tuff cone and the sea gradually eating into the core. *Reproduced from Charles Darwin,* Geological Observations on the Volcanic Islands and Parts of South America Visited During the Voyage of H.M.S. Beagle *(New York: D. Appleton and Company, 1891)*.

FIGURE 1.3. ***El León Dormido, or The Sleeping Lion, just off the north shore of San Cristóbal is a classic example of weathering.*** Darwin guessed that the amorphous mass of rock once filled the central hollow of a cone, the sloping walls of which have long since worn away. *Reproduced from Charles Darwin,* Geological Observations on the Volcanic Islands and Parts of South America Visited During the Voyage of H.M.S. Beagle *(New York: D. Appleton and Company, 1891)*.

Fissures of Eruption

Most of Darwin's explanations for specific geological features hold pretty true. But he went further still, hoping to understand the very underpinnings of the archipelago as a whole. With his living quarters in a corner of the *Beagle*'s chart room, watching on as the ship's surveyors got on with mapping out the Galápagos as never before, he was perfectly placed to do so.

With the advantage of satellite imaging, it's easy to see that Isabela, the largest island in the archipelago, is in fact made up of six neighbouring volcanoes connected by the extent of their laval outpourings. By looking at the *Beagle* map, Darwin could see this too and reckoned the archipelago's volcanoes were sitting on a series of parallel lines running diagonally from the north-west to the south-east. This can be most clearly seen on Isabela, where three large volcanoes (now called Wolf, Darwin and Alcedo) are perfectly aligned. With a ruler to hand, it is also possible to draw a parallel line between Darwin Island in the north-west and Española in the south-east, taking in Wolf Island (as opposed to Wolf Volcano on Isabela), Santiago, Santa Cruz and Santa Fé along the way. In addition, Darwin reckoned that Fernandina and Isabela's Sierra Negra sat on a third parallel, with Pinta, Marchena and San Cristóbal forming a 'less regular fourth line'.

Darwin also fancied he could see another set of lines perpendicular to the first, linking Floreana and San Cristóbal, for instance, or Cerro Azul, Sierra Negra and Santa Cruz. These interesting patterns, Darwin speculated, might be explained by the existence of rifts on the ocean floor from which lava had blurted and islands had formed. 'The principal craters appear to lie on the points, where two sets of fissures cross each other,' he wrote.

The origins of the Galápagos Islands turn out to be a little more complicated than this, but in proposing 'fissures of eruption' beneath the waves, Darwin was certainly way ahead of his time. Even a century later, there was considerable opposition to the idea that the earth's surface was made up of tectonic plates, vast slabs of crust jostling for position as if part of some poorly fitting jigsaw. We have now mapped out these plates in minute detail and are even able to measure their annual drift to within millimetres. The existence of tectonic plates, pushing and pulling in different directions, helps account for so many of the earth's bizarre geological phenomena, including the Galápagos.

These islands, we now know, sit near the junction of three major tectonic plates: the Pacific Plate to the west, the Cocos Plate to the north and the Nazca Plate to the south (see Appendix C, Figure 2). At the faults between these plates, magma bubbles up from the earth's mantle, cooling to form new crust and driving the three plates in different directions. At the point where the Pacific Plate meets the Cocos and Nazca Plates, we get the East Pacific Rise; where Cocos meets Nazca we have the Galápagos Rift. But the Galápagos needed more than a couple of rifts spewing out lava to come into existence. Indeed, it would not exist at all were it not for the reckoning of another geological force.

The Hotspot

In 1963, a Canadian geophysicist by the name of Tuzo Wilson published a rather brilliant paper that proposed a new explanation for the origin of the Hawaiian Archipelago. Like the Galápagos, the Hawaiian Islands seem to sit on a straight line, from Big Island in the south-east to Ni'ihau and Kaua'i in the north-west. Until the point that Wilson penned his manuscript, the conventional explanation for this alignment, put forward as recently as 1960, was similar to the one that Darwin proposed for the Galápagos: the Hawaiian Ridge had to be a result of 'a major fracture in the Earth's crust through which lava has poured at different centers'.

Wilson had good reason to think this was not how things had happened. There was not the slightest evidence for a fault beneath Hawaii— or, for that matter, beneath any of the other island chains across the Pacific. Even if there were such much mooted cracks in the ocean floor, they could not explain why Hawaii aged from east to west, its islands lined up like schoolchildren in a playground, with the oldest at the western extreme and the youngest, most active volcano to the east.

Wilson had a much better idea. In his paper, he did not refer to 'hotspots' but talked of 'sources of lava' originating deep within the earth's core. If the Pacific Plate were moving steadily, rather like a conveyor belt, in a north-westerly direction over such a powerful source, it would explain the extraordinarily linear arrangement of the islands.

'Each volcano, as it was carried away from its source, slowly became inactive,' he envisaged. 'The farther a volcano is from the East Pacific Rise, the older it is. The longer a chain, the older is the chain.' Rather wonderfully, the idea of a single, powerful hotspot also explains why Hawaii doesn't really stop at Ni'ihau and Kaua'i. In fact, it would be best to think of the Hawaiian Islands as the exposed peaks of a vast submarine mountain range, the Emperor Seamount Chain, that extends for thousands of kilometres into the Pacific.

The Galápagos Islands, we now know, are fired up by one of Wilson's 'sources of lava', a deep-seated hotspot that periodically sends volcanoes bubbling to the surface of the Nazca Plate. In contrast to the Pacific Plate (on which Hawaii sits), this slab of crust is conveying these eruptions off to the east rather than the west (courtesy of the combined forces of the East Pacific Rise and the Galápagos Rift). So, as in Hawaii, the islands we call the Galápagos are just the manifest pinnacles of a much longer mountain range known as the Carnegie Ridge. Of the volcanoes that do still peak through the waves, those closest to the hotspot in the west are the youngest, highest and most active (like Fernandina and Isabela's Cerro Azul, whose oldest lava flows date to some 500,000 years ago), and those furthest from the hotspot in the east are the oldest, lowest and least active (San Cristóbal and Española, appearing somewhere between 3 and 4 million years ago). See Appendix C, Figure 3.

Interestingly, there is another submerged mountain range, the Cocos Ridge, that stretches from the Galápagos region to the north-east all the way to Panama. The existence of this second blurted sequence of volcanoes suggests that the Galápagos hotspot once sat closer to the Galápagos Rift and perhaps even beneath the Cocos Plate itself.

It also suggests that at their eruptive genesis, these now submerged mountains probably broke through the surface just as Fernandina does today. As they were carried away from the hotspot, they cooled, contracted, eroded and returned beneath the waves, no longer bona fide, above board islands but submarine ones known as 'seamounts'. Studies that have taken a closer look at the Carnegie and Cocos Ridges have found their geology to be similar to that of today's Galápagos, indicating they were probably borne from the same hotspot. Other work has even found evidence to support the idea they were once real islands. Many of the seamounts have flat tops, most likely smoothed by the weathering action of wind and rain. Others are covered with cobble-like stones thought to have been caused by the gentle lapping of waves. Some contain mollusc-bearing limestone, deposits that are probably the remains of beaches long since lost.

This is more than just an interesting aside. It means that the islands we have come to know as the Galápagos will, one day, recede beneath the waves. First Española, then San Cristóbal and so on from east to west, each assuming a monumental position on the undersea Carnegie Ridge. But it's unlikely the Galápagos flora and fauna will sink with them. For the hotspot will, by then, have churned out new islands, naked slopes eager to be clothed by the plants and animals on the islands to their east.

As we shall see, this process of succession is crucial to the understanding of the Galápagos. In many instances, the ancestors of modern Galápagos species have been island hopping for millions of years. The current islands are merely the temporary homes of these highly evolved and much prized species. In the future, their descendants will have colonised islands that are, as yet, but a twinkling in the Galápagos hotspot's fire-filled eye.

Chapter 2. Ocean

Darwin Bay on the north-easterly island of Genovesa is a wonderful place to snorkel. It is also the site of the first dive in Galápagos waters back in 1925. On 9 April, gung-ho American naturalist and explorer William Beebe sent for his diving helmet. It was a classic bit of kit, a cumbersome copper affair replete with reinforced glass and rivets that left Beebe looking like a cross between a medieval knight and a *Doctor Who* cyberman. He plugged a hose ('of the common or garden variety') into the right ear of the helmet, through which one of his colleagues at the surface pumped a steady supply of air.

Beebe descended slowly to a depth of around 5m and took in his surroundings. At first, it all looked much as things had done during practice sessions he'd conducted at the New York Aquarium. So he sat down, Zen-like, on a convenient rock, shut his eyes and took a moment to reflect. 'I am not at home, nor near any city or people; I am far out in the

FIGURE 2.1. **William Beebe.** The American naturalist and explorer was the first person to dive in the Galápagos in 1925. His helmet—'a big, conical affair of copper'— weighed almost 30 kg. *Reproduced from William Beebe,* The Arcturus Adventure *(New York: Putnam, 1926).*

Pacific on a desert island, sitting on the bottom of the ocean; I am deep down under the water in a place where no human being has ever been before,' he wrote in *The Arcturus Adventure*.

When he opened his eyes again, Beebe found himself staring at a bizarre fish, 'the strangest little blenny in the world, five inches long and mostly all head'. Its long snout, flaring nostrils and two horn-like struc- tures that curved from the top of its head made it look absurd, almost like a prize bull. 'My blenny's eyes were silver with hieroglyphics of purple in them, and as I looked, he puffed a puff of water at my window and was gone.'

Beebe made five rapturous dives that day. In spite of all his preparation, he was nervous, especially during his second descent. Treading water beside the flat-bottomed research boat, just about to duck into the helmet to commence the dive, he saw a huge shark, 'a giant of a generous eleven or twelve feet, cutting the water with his great dark fin'. The sight would have been enough to send most people scrambling back up the ladder and into the boat, but this was not an option. Beebe had been forever telling his colleagues that sharks are harmless; he could hardly bottle out now.

On the bottom, with the shark still on his mind, he found the restricted field of vision imposed by the helmet somewhat alarming. 'I am certain that from above I must have looked like some strange sort of owl, whose head continually revolved first in one and then in the opposite direction,' he wrote. To make matters worse, he found the fish around him incredibly curious. Triggerfish are notorious in this respect, their colourful, rhomboid bodies forgoing their usual preoccupation with crustaceans to take a nip at a passing diver or snorkeler. 'I would often leap up in expectation of seeing some monster of the deep about to attack me,' wrote Beebe.

Once he'd calmed down, he was able to take in his surroundings. The sun's rays filtered down from the surface 'as though through the most marvellous cathedral'. He saw white-striped angelfish chasing one another 'in sheer play'; brightly coloured wrasse 'slender and supple as eels'; a mist of yellow-tailed surgeonfish passing before him; an octopus flowing over a rock 'like some horrid viscid fluid in animal form'. Fairly limited in his movements owing to his weighty diving clobber, Beebe soon hit upon the idea of waving a freshly killed crab through the water to bring the fish to him. Within minutes he was surrounded by hundreds of specimens from dozens of different species. 'Their keen powers of scent drew them like filings to a magnet,' he wrote. 'Often there was a central nucleus a foot or more in diameter, of solid fish, so that the bait and my arm to the elbow were quite invisible.'

The Deep

Impressive as this kind of experience might be, it's worth bearing in mind that it is necessarily superficial. Far deeper down than it's possible to dive with scuba gear, there exists an entire world that few humans have ever set eyes on. In the 1970s, marine geologists got the first glimpse of it in the waters around the Galápagos, when they dropped a probe overboard and lowered it to a depth of several thousand metres into the vicinity of the Galápagos Rift (the east-west cleft between the Cocos Plate to the north and the Nazca Plate to the south, lying some 250 km north of Puerto Ayora on Santa Cruz). They were looking for signs of so-called hydrothermal vents discharging plumes of superhot water from the sea floor. Photographs and temperature measurements taken by the probe produced the first clear evidence of these chimney-like structures that we now know are a feature of all the world's oceans. One of the photographs revealed something even more surprising. It is grainy, but one can still make out a clump of giant mussels, each about 15 cm long. A little way off, a crab scuttles across the lava bed.

In 1977, the US National Oceanic and Atmospheric Administration sent the deep-sea submersible *Alvin* through the Panama Canal to take two men down for a closer look. In February and March, they made twenty-four dives in the region where the 1976 expedition had produced its photos. The warm, chemical soup spewing from these vents seemed sufficient to sustain an entire community of weird creatures in total darkness at extraordinary pressure. When *Alvin* focused its beams on the craggy lava floor of the Galápagos Rift, it shone light on a monstrous abundance of giant mussels, crabs, oysters, limpets and tubeworms growing up to 3m long.

Floating in the huge water column above, many more species have yet to be seen, let alone described. Beebe caught a few of these, like a lanternfish netted en route from Genovesa to Isabela. Bringing it into the dark room on board the research vessel *Arcturus*, he marvelled at the 'little eruptions of body fires which flashed forth' from its flesh. Dredging off Fernandina, Beebe's team pulled up a couple of anglerfish with their weird lamp to illuminate the pitch before them. The BBC's 2006 three-part documentary *Galápagos* provides an even more impressive glimpse

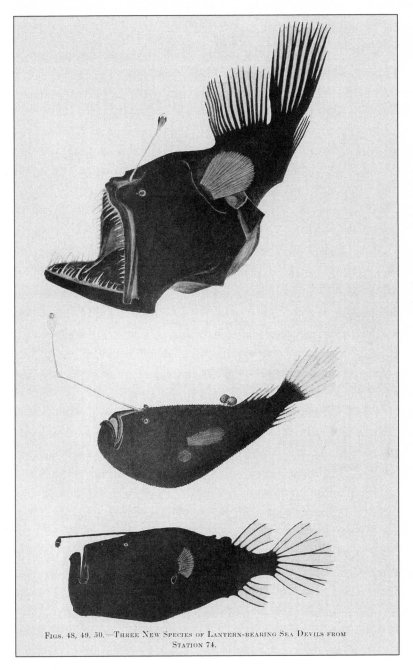

Figs. 48, 49, 50.—Three New Species of Lantern-bearing Sea Devils from Station 74.

FIGURE 2.2. *Three new species of lantern-bearing anglerfish or 'sea devils'.* This was the kind of extraordinary life that Beebe's team brought up from the deep, dark water to the northeast of the Galápagos. *Reproduced from William Beebe,* The Arcturus Adventure *(New York: Putnam, 1926).*

into this unknown world when one of the cameramen braves a night dive far out at sea to film the bizarre luminous creatures that drift up to the surface when it's dark. As the Galápagos National Park Service does not normally permit night diving, only through films like this can we see this other world. But it's exciting to think it's out there.

Currents

Of course Charles Darwin saw none of this hidden world. His total fish haul in 1835 was just fifteen species, netted from the shallow waters around San Cristóbal and Floreana and preserved in a cask of wine. There were a few interesting species amongst them: the Galápagos sheepshead wrasse with 'four very conspicuous, strong, curved, canine teeth' sticking out from the front of each jaw; the tiny red clingfish, which looks like it might have been hiding away in a crevice since the Cambrian era; and the bullseye puffer, which surprised Darwin by making a 'loud grating noise'. Although Darwin's fish collection from the Galápagos was not much to write home about, it nevertheless hints at the amazing diversity: all the species he collected were new to science, 'without exception'.

It is also clear from Darwin's diary just how rammed with life the Galápagos waters used to be. 'Fish, Shark & Turtles were popping their heads up in all parts,' he noted at an anchorage off the west coast of San Cristóbal. The Beagle's crew eagerly flung fishing lines overboard and quickly hauled in 'great numbers of fish', many of them huge at almost 1m long. 'This sport makes all hands very merry; loud laughter & the heavy flapping of the fish are heard on every side,' he wrote.

This productivity is, in large part, down to the cool, nutrient-rich waters that reach the surface in the Galápagos. The Beagle's captain, Robert FitzRoy, was fascinated by the islands' 'very remarkable' currents, noting their speed, persistent direction and also the 'surprising difference in the temperature of bodies of water moving within a few miles of each other'. To the east of Isabela, for instance, FitzRoy found the surface water to be a comfortable 26°C. On the island's western shores, by contrast, the sea temperature was over 10°C lower.

This striking difference is a result of the currents that meet in the Galápagos (see Appendix C, Figure 4). The strong, cold Humboldt

Current flows up the coast of Peru, combining with the weaker, warmer Panama Current from the north to form the South Equatorial Current, which surges out into the Pacific from east to west along the equator. The Cromwell Current (or Pacific Equatorial Undercurrent) runs in the opposite direction, a deep course of even colder water coming in from the west. When this hits the western edge of the Galápagos, it has nowhere to go but up, bringing its especially chilly water to the surface around Fernandina and the west coast of Isabela.

These currents have a major influence on the peculiar climate in the Galápagos. Between June and November, the Humboldt Current is the major player, its cold, north-westerly influence turning down the temperature in the islands. During this cool season, there is precious little rain along the coast, but at higher altitudes any moisture in the warmer air tends to condense into a drizzling mist, or *garúa*. When the Humboldt slackens in December, the Panama Current becomes ascendant, the mist evaporates from the highlands and the hot season begins. As the temperature of the water rises, evaporation increases, clouds grow and rain falls. In an El Niño year, when the Panama Current is particularly warm, the Galápagos can experience a deluge.

Beebe may have encountered what we'd now consider to be an El Niño. 'Rain in the afternoon and showers in the evening,' he recorded in the *Arcturus* logbook on 8 April 1925, the day before he tried out his diving helmet for the first time. Although he had been to the Galápagos once before, at around the same time of year in 1923, the rain in 1925 seemed exceptional. A couple of weeks later, on Española, Beebe noted 'small greenish rain pools deep among the rocks' and a freshwater pond 'a half-mile in length'.

The conditions in 1924–1925, however, were nothing like those experienced in 1982–1983, when more than 3,500 mm of rain fell on Puerto Ayora, roughly ten times more than normal. High sea levels, heavy swells and turbulent waters caused widespread erosion; 'stupendous thunderstorms' resulted in flash flooding and human fatality. In 1935, Norwegian settler Alf Kastdalen had been one of the first to make a go of life in the Santa Cruz highlands. As a farmer, he had a special interest in the weather, noting how a creek near the family home would flow regularly in some years and be bone dry in others. In 1983,

Kastdalen—then in his sixties—was electrocuted as he tried to replace a high-tension cable torn down by the El Niño elements.

Inshore

In addition to affecting the climate, the currents obviously also have a major influence on what goes on beneath the water. When the Humboldt and Cromwell Currents are dominant, they bring nutrients lapping up against the Galápagos shoreline, feeding blooms of plankton and allowing sun-loving algae and grasses to bed down on the pillows of lava beneath the surface. These two currents are also rich in oxygen, which makes life a lot easier for filter feeders like sponges and corals, invertebrates such as sea stars and urchins, and almost five hundred species of fish.

Some of the most productive waters are in the Bolivar Channel between Fernandina and Isabela (where Captain Morrell witnessed the violent eruption in 1825). Beebe noted the extraordinary diversity and abundance of species when, diving off the north coast of Fernandina, he found himself 'amidst sea-weed so tall and thick it was like a corn-field'. A gang of giant sea bass ('too ugly and dangerous to call a school') came mooching through the waving weeds, all of them olive and brown, 'their ugly jaws chewing eternally on the cud of life'. Then Beebe caught sight of a glimmer of gold in their midst, a fish that seemed to be part of the shoal, nudging its neighbour and getting nudged back in return.

Darwin collected this species, though as he only netted a 'mottled brown' specimen he could not have guessed, as Beebe did, that it comes in different colours. The bacalao (as it's known) is the largest sea bass living in the Galápagos, with some specimens weighing more than a human toddler and around one in twenty coming in the mysterious golden form. Other fish families have similarly confusing morphs, notably the hogfishes and wrasses, several species of which can vary from the drab to the distinctly psychedelic. One of the largest in the Galápagos is that sheepshead wrasse that Darwin saw (sometimes more than 70 cm long), which varies from red to bluish-grey (though a blazing spot above its pectoral fin remains a constant gold). The guineafowl puffer too comes in a couple of different forms: either in black with white polka dots (like a guineafowl) or in a stunning golden form. As yet, nobody has come up with a satisfactory explanation for why one species should come in different colours, but more than likely it has something to do with sex.

The bacalao and other sea bass, the hogfish, the wrasse and the guineafowl puffer all turn out to be hermaphrodites. This means they're able to produce both eggs and sperm, though probably not at the same time. Most of these species start out life as females, with only their egg-producing tissue active. Then at some stage, triggered by as yet unknown cues, the females undergo a sex change and switch to producing sperm instead. In some cases, it's possible that this radical transformation is accompanied by spectacular changes in colour.

The fact that we don't really know the reproductive set-up for most of these bizarre fish is of more than just academic interest. If we don't know how they breed, we can't anticipate how the population will fare when faced with pressures from fishing or climate change. This is of particular concern for a species like the bacalao, one of the most commercially valuable fish in the Galápagos. It is thought that in the 1980s this one species made up as much as 40 percent of the value of the total catch. It remains sought-after today, particularly the rare golden form, though over the last few decades Galápagos fishermen have turned their attentions to the even more lucrative spiny lobsters and sea cucumbers, also at their most abundant in the Bolivar Channel.

Before the Chinese began to pay big bucks for sea cucumbers, nobody thought to fish them from Galápagos waters, and there would have been thousands lying on the sea floor off an island like Fernandina. By the time of the first official census in 2001, an area the size of a football pitch held just 115 specimens of the Galápagos sea cucumber, the species that is commercially exploited. Today, the same area is home to fewer than five. For a creature that relies on sperm and egg finding each other in the water column, such ridiculously low densities spell almost certain extinction.

The abundant plant life attracts turtles too. There are occasionally hawksbills, but the Pacific green turtle is by far the most common species in the Galápagos. It dines on a diet of algae and seagrass, supplemented by the occasional mouthful of mangrove or bite of crustacean, so tends to hang out in protected bays, inlets and lagoons. The females lay eggs between December and June, returning in the cover of darkness to the beach on which they hatched to carve a pit-like nest chamber in the sand. They can drop as many as eighty golf ball–sized eggs into the nest, before covering them with sand and returning to the sea, a feat that a female may repeat several times over the course of the season. Between six and ten weeks later, the hatchlings emerge. Most of them do this at night, when there are fewer predators about, but many still succumb to the eager ghost crabs and yellow-crowned night heron that line up in wait.

Fur Seals and Sea Lions

The cooling, sustaining effects of the Humboldt and Cromwell Currents also make the Galápagos habitable for fur seals and sea lions, creatures more often associated with ice than with the equator. But it's a perilous existence. When the life-giving currents stall, a devastating chain of events ripples through the Galápagos ecosystem. First to go is the microscopic plant and animal life, followed swiftly by the fish that feed on them, and so on, all the way to the fur seals and sea lions near the top of the food chain. During the particularly harsh El Niño of 1982–1983, for instance, almost every fur seal pup under the age of three years died.

In spite of the frequency of this devastating collapse in the productivity of Galápagos waters, the fur seal and sea lion appear to have survived here for hundreds of thousands, if not millions, of years. But by surveying a map of where these species prefer to hang out, it's clear that the fur seals favour the west of the archipelago where they have access to the resource-rich Cromwell Current. Sea lions, by contrast, which are larger and so can dive to greater depths, are able to survive throughout the archipelago.

Owing to the frequent fluctuations in the marine environment, these marine mammals all show considerable flexibility in their behaviour, particularly when it comes to reproduction. They must be prepared to mate whenever the conditions are right. This need to breed at short notice places a particular burden on bull fur seals and sea lions, which cannot afford to let up in their defence of a territory. They put so much effort into keeping their harem of females all to themselves, they hardly have time to eat. A sudden shortage of food can be fatal. In the 1982–1983 El

Niño, for example, almost all territorial bull fur seals and sea lions disappeared, presumed dead.

Even with all the effort they put in, bull sea lions in the Galápagos don't get as much paternity as they might hope for. In the case of the elephant seal of Antarctica, where dominant males are colossal and only need to fight off other males for a few crucial days when females are fertile, they are able to do most of the mating and father most of the offspring. In the Galápagos, this is not the case. A bull sea lion simply can't rebuff the advances that smaller males continuously make on the females in his harem, and quite a few of these others end up as fathers too.

Females don't have it easy either. Compared with their closest relatives that inhabit the poles and can be weaned at just four months, fur seal and sea lion mothers in the Galápagos must provide milk for their offspring for up to three years. This means that if they give birth the following year (as they sometimes do), they might have a yearling and a newborn pup both expecting milk. Unless the conditions are really good, it just may not be possible to juggle two youngsters like this, and one of them will die.

Dolphins and Whales

Much as they were abundant with fish in Darwin's day, the Galápagos waters were once also flush with dolphins and whales. These creatures—principally the sperm whale—led to the first commercial exploitation of the archipelago. In April 1794, James Colnett, captain of the British whaling vessel HMS *Rattler*, elaborated on the abundance of the creatures. Cruising to the west of the islands, his crew encountered sperm whales 'in great numbers,' he wrote of his whale-scouting expedition a few years later. The more experienced whale hands on board 'uniformly declared that they had never seen spermaceti whales in a state of copulation.' There were also newborns, which he judged were about the size of small porpoises. 'I am disposed to believe that we were now at the general rendezvous of the spermaceti whales from the coasts of Mexico, Peru, and the Gulf of Panama, who come here to calve,' he wrote. It was an astute observation. For some reason, as yet unknown, females and their young remain in Galápagos waters throughout the year, whereas the far larger males only make the occasional appearance to mate. For the whaling fleets that subsequently descended on the Galápagos, these sperm whales were big business. 'The situation I recommend to all cruizers,' wrote Colnett.

Today, the descendants of these and other cetaceans (the collective name for porpoises, dolphins and whales) are no longer harpooned but revered, with more than a dozen dolphin species and about as many whale species sighted in the Galápagos Marine Reserve. As the whaling fleets that followed in Colnett's wake discovered, the best place for a cetacean encounter is to the west of the islands. It's here that tourists are most likely to find bottlenose dolphins riding the bow waves of their vessels, to glimpse a pod of orcas or even to see the flukes of a blue whale in the distance. But the actions of Colnett and his fellow whalers mean that such sightings are relatively rare, even today.

Rays and Sharks

By contrast, rays and sharks are far easier to spot. Snorkelers and divers will frequently encounter stingrays, spotted eagle rays, golden rays and sometimes even the vast undulating manta ray. The whitetip reef shark, a wide-ranging species with a telltale daub of creamy white on the tip of its dorsal and caudal fins, is still fairly common in both the Pacific and Indian Oceans. The Galápagos shark, in spite of its name, is not unique to the archipelago; it was just discovered here. 'We examined a large number of them, several hundred being taken aboard the schooner, and we saw probably thousands in the water,' wrote the scientists of the California Academy of Sciences in 1905.

They found the Galápagos sharks most abundant around the north-westerly outlying islands of Darwin and Wolf. This is the best place to see scalloped hammerhead sharks, particularly just to the southeast of Darwin, where they congregate in great numbers during the daytime. It remains a bit of a mystery why they do this. It could be that the currents are favourable, allowing the sharks to rest up before they head out at dusk to forage. Perhaps this schooling fulfils some as yet unknown social need. They could be coming together to attract the services of smaller fish that will pick parasites from their skin. Darwin Island could just be conveniently close to the shoals of smaller fish on which they feed. Whatever the reason, congregations of scalloped hammerheads and smooth hammerheads, as well as the chance of seeing a colossal whale shark, draw a lot of visitors to this corner of the archipelago. The six dive sites dotted around the islands of Darwin and Wolf are the most highly sought of seventy-five dedicated marine visitor sites in the Galápagos and crucial to the buoyancy of the island's dive tourism industry, estimated to be worth some $20 million a year.

Remarkably, William Beebe seems to have anticipated the interest that people would have in diving. 'I am deep down under the water in a place where no human being has ever been before; it is one of the greatest moments of my whole life,' he wrote. And then, in a throwaway adjunct, he reckoned that 'thousands of people would pay large sums, would forego much for five minutes of this!'

How right he was.

Chapter 3. Seabirds

With the waters around the Galápagos offering such riches, it is not surprising that the islands are home to a great abundance of seabirds, and most famous amongst these are the boobies.

Boobies

The blue-footed booby is a real character and certainly one of the most colourful Galápagos birds. Part of its appeal is its unquestionably clown-like appearance. Yet, in spite of its googly eyes and wondrously coloured, apparently oversized feet, at sea it is transformed. When a squadron of a hundred or so blue-footed boobies homes in on a shoal of anchovies, each bird hauls in its wings, faltering for a moment in mid-air before pivoting to face the ocean, and then plummets. The speed and unwavering direction are impressive. The splashless entry into the water, as after a perfectly executed Olympic dive, has the effect of breaking up the shoal.

As the booby swims back to the surface, it's relatively easy to pick off isolated fish silhouetted against the bright sky.

Between these extremes of comedy and anatomical perfection, there's been a fair bit of fascinating research into the breeding ecology of these birds, and their striking feet are planted firmly at the centre of the story. When it's time to breed (between June and August), males and females forge a bond based on the blueness of their feet. As with most other birds—think peacocks—it's the male that's particularly showy. Not only are his feet a deeper, brighter blue than the female's, but he makes a special effort to show them off, flying into his territory with his pedal extremities splayed in contrast against his bright white under-belly. Once he's done enough to attract a female, the putative pair will engage in an exaggerated foot-raising exercise, each bird flapping from one webbed foot to the other. As things get really serious, the boobies go in for 'sky-pointing', when one of the pair (usually the male) stretches its neck and bill towards the sky, extends its wings to the sides and lets out an evocative cry, rather like a mournful blast on a Peruvian panpipe. The birds will also bring each other twiggy little offerings in a symbolic nest-building ritual, swinging the gift up and around before placing it near the other bird's feet.

Why so much emphasis on feet? A very neat and simple experiment gives us some clues. Cross-fostering is a standard trick used by zoologists in an effort to disentangle the influence of genetic and environmental factors on phenotype (a catch-all term that translates, roughly, as 'what an individual is like'). In 2002, researchers working with blue-footed boobies on an island off Mexico created a colony of unwitting foster parents. When an egg hatched in one nest, they swiped the chick and swapped it with another of a similar age. So all the booby pairs they toyed with had the same number of chicks; it's just that one of them was not their own. This set-up allowed the researchers to demonstrate that the colour of the caregiving male's feet seems to have an influence on chick condition: the gaudier the feet of the foster father, the faster and fatter his chicks will grow.

The most likely explanation for this is that the blueness is an honest indicator of fitness, a badge of quality, an unforgeable certification of health. What does foot colour reveal? At least a couple of things. First, it communicates something about the male's nutritional status. We know this because if a male blue-footed booby is deprived of food for just forty-eight hours, the colour will rapidly drain from his feet. This is because he's no longer getting the natural carotenoid pigments from his diet that he uses to gloss his feet. Give him back his food, and the colour will flush back into them in a matter of hours. Second, foot colour may also reveal something about health. When researchers injected males with a low-level challenge to their immune system, their feet got progressively duller over the course of the infection. The reason for this is that carotenoids play a role in stimulating the immune system. Only males in fine fettle can afford to channel these valuable compounds into their feet.

If you're a female blue-footed booby, then it would make sense to go for the most colourful male you can. He is feeding well and disease-free, he may pass on some of his innate hunting talents or immune strength to your chicks and he is likely to provide well for them as they grow. The researchers confirmed that females pay attention to this trait with another nifty little trick: just after the female had laid her first egg, they took makeup (a 'non-toxic and water-resistant blue') to the feet of half the males in their sample, turning their feet from vivid aquamarine (the most attractive) to a pale blue (considerably less desirable). The other males acted as controls.

The faded males didn't seem to notice their change in status but continued to display and court their partners much as they had been. The females, however, were not so ambivalent. They suddenly reduced the frequency of sky-pointing, symbolic nest building and sex. That's not all. When females paired with these painted males came to lay their second eggs, they were of significantly smaller volume with significantly smaller yolks than those laid by partners of the controls. The conclusion is clear: a female booby will keep a beady eye on the colour of her partner's feet, using it to constantly reassess the value of their relationship and her investment in it.

But is this just about females choosing males? Before we move on from blue-foots to consider the two other booby species in the Galápagos, we need to address the fact that female boobies also have blue feet (though admittedly not quite as garish as those of males). In the same way that nipples are a developmental—if redundant—exigency in male mammals, so the biochemical shenanigans that male boobies use to convert dietary carotenoids into bluish pigments might just be something females must put up with. Alternatively, males might be exerting some kind of choosiness themselves. Why not? In most bird species, females are notable for their drab appearance and absence of elaborate sexually selected traits. But in those species where males contribute substantial amounts to the breeding effort—as in the blue-footed booby, for instance—one might expect males to be a bit more discerning about whom they mate with.

In a follow-up study that showed an admirable absence of gender bias, the booby researchers sought to find out, repeating the makeup experiment but this time painting the feet of females rather than males. Lo and behold, males pay attention to a female's foot colour too. Dulled-down females received much less attention from their partners—and indeed from other males in the colony. It's all in the feet. If they are bright blue, verging on green, the bird is likely to be happy, healthy and attractive. If they are pale blue, almost white, then it's struggling.

The Nazca booby is a less comical spectacle, an altogether snazzier bird with a white glossy body and a Zorro-like mask around its eyes. Is it hiding something? Those ornithologists who've taken a good look at its behaviour have discovered it is. Nazca boobies lay two eggs, the second as insurance against the failure of the first. But when both hatch, the first chick to do so—let's call her A chick—wastes no time in attacking her younger sibling, forcefully ejecting B chick from the nest. Because the Nazca female lays its eggs several days apart, A chick has a distinct size advantage, and the greater that advantage, the quicker A chick is to expel its diminutive sibling. Unfed and unprotected, B chick faces certain death, its corpse destined to be hoovered up by a hungry frigatebird or other scavenging beast.

This is not all. Nazca booby society holds an even darker secret. For some reason, juvenile females suffer higher mortality than males. The result is there are many more adult males than females. Those males that can't find a partner get up to something really nasty, wandering through a colony in broad daylight in search of unguarded Nazca nestlings (a fairly common sight once they reach around one month of age and their parents begin to go on extended foraging trips). In around one-third of instances, the unemployed adult will behave rather nicely to the chick, standing beside it, preening it or even presenting it with gifts of twigs, pebbles or feathers. More often, though, the adult will abuse the young-ster, jabbing at it with its bill, plucking out its down or biting it and shak-ing. Occasionally, these advances are sexual too, the aggressor climbing onto the chick's back and squashing it beneath its feet as it attempts to mate.

Apart from the emotional trauma caused by this abuse (at which we can only guess), the nestlings may suffer an injury and, as a result, can die. A red scratch or gaping wound is hard to conceal against such bril-liant white fluff, and mockingbirds will peck at a nestling's wound to top up their diet with some nutritious Nazca blood. Perhaps the most startling finding is that those boobies bullied as chicks are more likely to go on to torment nestlings as adults. In an extraordinary paper pub-lished in 2011, the Nazca researchers referred to a 'cycle of violence', a phrase usually reserved for discussions of child abuse in humans. Their findings, they concluded, 'provide the first evidence from a nonhuman of socially transmitted maltreatment directed toward unrelated young in the wild'. When I first visited the Galápagos in 2003, I found the Nazca more attractive than the blue-footed booby. Now I'm not so sure.

In contrast to the blue-footed and Nazca boobies (which are only found in the eastern Pacific), the red-footed booby can be found throughout the world's major oceans. But in the Galápagos it has a pecu-liarly limited distribution. This may have something to do with its food of choice—flying fish—which means it's sensible to set up home on the peripheral islands of Genovesa, Wolf and Darwin with good access to the deep sea. However, there may be another reason too. The red-foot's foraging trips can take parents away from the nest for long periods, with nestlings typically left unattended for over ten hours every day. Being so vulnerable to the depredations of the Galápagos hawk, the red-footed booby has only been able to establish itself on islands where the hawk is absent. Indeed, the disappearance of hawks from San Cristóbal and Floreana (owing to the human presence there) seems to have allowed red-footed boobies to establish new colonies on these islands.

Juvenile red-foots appear to be particularly playful, sometimes gath-ering in large flocks of around one hundred birds and engaging in what's been described as 'a curious communal game'. Working on Wolf in the 1980s, a pair of ornithologists observed the booby game on several occasions and described its rules: 'The game, if we may call it so, always started near the summit of the island.' From their observation point the scientists could not quite see what the birds were playing with ('it might be a feather, a leaf or a stick'), but one individual would kick off proceedings by dropping the 'toy' into the cloud of birds below. 'After a few seconds of free fall it is caught by a second bird and carried upwards, only to be dropped again and subsequently picked up by yet another bird.' The game would finish when the wind had carried the flock out to sea and the toy had fallen onto the surface. On the same expedition, the ornithologists also noted several peculiar objects ('3 short plastic thongs, 1 toy soldier and 1 toy bassoon') beneath the nesting and roosting spots of the red-footed boobies. 'Are these the boobies' "toys"?' they wondered. 'How else can these strange objects make their way to the top of Wolf Island?' How indeed?

Galápagos red-footed boobies are also notable for the fact that most individuals are brown (apart, of course, from their crimson boots, their blue bill infused with pink and the inky stain around their eyes). A quick survey of populations elsewhere reveals that white-feathered red-foots are by far the most common, making up around 95 percent of the global population. But there are also a few populations—as in the Galápagos—where brown is the dominant hue.

This is the sort of pattern that would have fascinated Darwin, and it has caused a good deal of head scratching amongst red-footed booby researchers. We now know that the brownness in Galápagos red-footed boobies is caused by a couple of mutations in a single gene—that encoding the melanocortin-1 receptor (small variations of which are known to cause dramatic changes in pigmentation). But is there something more interesting going on here? Is it possible that brown boobies are at some kind of advantage over white boobies in the Galápagos? Perhaps the darker feathers offer some kind of thermoregulatory benefit? Maybe brownness plays an important role in pairing, rather like the blueness of the blue-footed booby's feet. Perchance it provides the birds with some kind of camouflage, helping them to sneak up on their prey. Or what if, as has been suggested for brownish red-foots that live on Europa (a small island in the Mozambique Channel between continental Africa and Madagascar, not the smallest of Jupiter's four Galilean moons), the brown makes them less susceptible to avian bullying by the likes of frigatebirds? We don't yet know.

Frigatebirds

In the Galápagos, there are two species of frigatebird: the great and the magnificent. These can be pretty hard to distinguish, though the great frigatebird is the smaller of the two, and its adults have a greenish tint to their feathers rather than the magnificent frigatebird's purple gloss. The youngsters, often perched patiently in a scrubby nest of twigs, are easier to tell apart: the chest and neck of a great frigatebird juvenile tends to be rust stained, whereas an immature magnificent has a more vulturous look, its white chest and head standing out against its otherwise dark plumage. Just why these youngsters take so long to reach full-blown adulthood (at least eight years for a great frigatebird female and around ten years for a male) remains a bit of a mystery. But the development of adult plumage appears to proceed in fits and starts, sometimes with no moult for up to four years. This suggests that perhaps these birds face an internal energy crisis, unable to gather sufficient food to afford the expense of refreshing their feathers. This is why many of the juveniles have a distinctly ragged, rather unsavoury look about them.

Both great and magnificent frigatebirds have huge wings relative to their body size. This might make for an awkward lolling gait and an ungainly, flapping take-off, but once airborne it allows them to stay aloft for long periods, cover great distances and reach speeds of around one hundred miles per hour. Their deeply forked tail gives extraordinary manoeuvrability, helping them to snatch at flying fish or scoop squid from the ocean's surface without entering the water (as they'd struggle to get airborne again). These aerial acrobatics also mean they can dabble in kleptoparasitism (thieving to you and me). A frigatebird often looks out to sea from its thorny perch, keeping a jet-black eye out for incoming traffic: red-billed tropicbirds, terns, boobies. When it spies a suitable target, it attacks. For the seabird returning from its feeding grounds, the sensible thing to do is to regurgitate a morsel of half-digested fish, which is usually enough to divert the attacker from its offensive so as to pluck its winnings out of the air. If nothing is forthcoming, however, the frigatebird will strike, its hooked bill grasping the victim's tail or wing to encourage it to release some of its catch. These assaults can be brutal, with both birds dropping out of the sky and the smaller bird coming away with damaged feathers or, worse, a broken wing.

During their troubled adolescent years, it can be rather hard to tell a male from a female frigatebird. But once a juvenile finally graduates to adulthood, it becomes a doddle to separate the sexes: both great and magnificent frigatebird females are marked out by their neat, understated white bib; the males go in for far more garish attire, an astonishingly crimson, turkey-like wattle they can inflate to the size of a football. Like the blue-footed booby's feet, this is clearly a signal that females pay attention to. This is especially obvious in a frigatebird colony during the breeding season, when males thrust out their wings, throw their heads to the sky and pump up their throats. It's not just the sight that's impressive; the sound is too. These courting birds begin to click, uttering a deep staccato that resonates inside their ballooning necks. The sight of dozens, sometimes hundreds, of such males showing off to females in a concentrated spot is reminiscent of what zoologists refer to as a lek, a mating ritual where females browse the wares on offer, mate with the most attractive male, then head off to rear the offspring as a single parent. But unusually for species with such ostentatious males, both great and magnificent frigatebirds form male-female partnerships, and males make a substantial contribution to the breeding effort.

Why, then, do males go to all the trouble of producing a fancy wattle and spend day upon day trying to impress the ladies? It probably has something to do with the balance of males and females. In one study of great frigatebirds on Tern Island in the middle of the Pacific, researchers found that amongst unpaired birds, there were typically two or three males (and sometimes up to ten) for every female. It remains unclear why this is so. Perhaps the sex ratio is somehow skewed at birth. Perhaps females are more likely to die. Or perhaps, with their greater overall investment in reproduction, not all females in a population can afford to breed every year. Whatever the reason, it looks like the extraordinary puffed-up courtship of the male frigatebird is a price worth paying for a chance of some seriously scarce action. It certainly seems to work.

Those males that can afford to do the most displaying have a significantly greater chance of ending up in a meaningful relationship with a member of the opposite sex.

So far, none of the seabirds we've looked at is peculiar to the Galápagos. Eagle-eyed as ever, Darwin made this observation himself, noting far higher levels of endemism amongst the land birds than the seabirds. The explanation is simple. As a general rule, seabirds travel further than land birds and tend to move in groups, so they will have found it relatively easy to reach the Galápagos and settle without much disturbance, as Darwin put it, to 'their mutual relations.' There are, however, several notable exceptions, seabirds that have now become completely tied to the Galápagos, found nowhere else on earth.

The Albatross

One of them is the waved albatross, a colossal bird with wings that span more than 2m from tip to tip. The only place visitors can reliably encounter this species is at Punta Suarez on Española, one of the most popular visitor sites in the Galápagos. From the dock at the eastern tip of the island, a trail winds up onto the eroded flat top of the island, which acts like a runway for the albatrosses, allowing them to build up speed before launching off the island's sheer southern perimeter. In fact, waved albatrosses have even colonised a bona fide but abandoned landing strip on the island, cleared during World War II to service a US radar base.

En route to the albatrosses, we'll just take a short digression to consider a couple other stunning (though not endemic) seabirds on show at Punta Suarez. Red-billed tropicbirds can be seen nesting in rocky crevices beside the tourist trail or cruising just off the island, their impressively long tail streamers steadying them against the prevailing south-easterly wind. It's also a good place to see swallow-tailed gulls, a beautiful bird with black tips to its wings, a slate-coloured head and what looks like a dab of white paint on its bill. They are also unusual amongst gulls for being nocturnal, a disposition discovered in the 1960s by a dedicated researcher who spent several nights sleeping rough on South Plaza (where there's also an active colony of these gulls). He set an alarm to wake him on the hour every hour so he could see what the gulls were up to. At dusk, he found, the birds began to leave the colony en masse so that by the time it was dark, the only individuals left on land were those incubating eggs or brooding chicks. By feeding at night the swallow-tailed gull may be sidestepping the competition, able to forage undisturbed by larger, more aggressive sea-birds like boobies, frigatebirds and albatrosses. They are also probably after particular prey species that come to the sea surface at night, as they prefer to forage during a full moon, when the abundance of these creatures is greatest. The swallow-tailed gull's bright-orange eye ring is wide, exposing a relatively large area of the eyeball to its environment and helping it to see in the dark.

Back to the endemic waved albatross. Between January and March, most of the birds are at sea. But from April onwards they begin to return, males first to stake out an area of clifftop and await the return of their long-term partners. These bonds between males and females are reaffirmed every year in a ritualized ceremony. Females lay a single egg between April and June but concentrated in May, usually on a flat patch of ground or beneath a bush. In the first few days of incubation, one or the other of the parents may move the egg, shuffling it along on top of its feet as emperor penguins do during the Antarctic winter. There's an obvious risk of breakage during such a manoeuvre, so there has to be a good reason why they do this. Nobody has really come up with a satisfactory one. Perhaps they've taken a dislike to a neighbour. Over the next two months, the male and female will share the burden of incubation, with one coming and the other going. This becomes more frequent as hatching approaches, with switching occurring every four days or so. Then an awkward chick emerges, with patchy brown down eventually puffing up to give the youngster a fluffy, daffy air. The parents will brood the young chick for several weeks, followed by a slightly more relaxed guard for a couple more, before they are prepared to leave it alone for days on end.

Owing to the size of the albatross, it's relatively easy to fit one with a satellite transmitter to find out where they go. When researchers did this in the 1990s, they found that the waved albatrosses head towards the coastal waters of mainland Ecuador and Peru. The signal from one bird was picked up more than 1,200 km from the Galápagos, at a latitude level with Lima.

the top of the island on an inspection tour. The country nearly all rocks, covered with dry vegetation—quite thick in places. Huge lizards were abundant, and a number were taken. We got two snakes, the largest about 3 ft. long. Birds not numerous, and many in very worn plumage. I saw *Certhidea*, mocking-birds, and three species of *Geospiza*. One hawk was seen. Doves very numerous. We

collected several black tern, and the sailors killed a yellow-crowned night heron (*Nyctanassa*). Beck brought back two goats from a flock of twenty that he saw.

Oct. 23.—Went ashore early after birds. Returned at 11 a.m. with about 50 small birds, two hawks, several oyster - catchers, a booby, sanderling, gull, and a

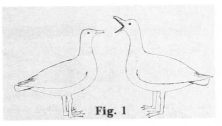

Fig. 1

lot of lizards ; also several black iguanas. We skinned all the birds before supper, with the exception of the two hawks, it getting too dark to work. In the evening several short-eared owls came out to see us, and I knocked one down with the spreader of the yawl-out.

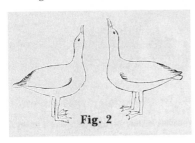

Fig. 2

Oct. 24 (Sunday).—Skinned a hawk before breakfast, after which Hull finished the other. I skinned the owl, spent another hour in fixing up things, and took a vacation the rest of the day. I fed one of the tortoises with banana peel, which it took from my hands.

Oct. 25—Shot 20 birds each, and returned alt 9:30 a.m. We skinned birds all the rest of the day. In the afternoon the mate and sailors went off on a goat hunt, but found no goats. They reported, however, a big albatross' rookery, and brought in several eggs of the albatross.

Oct. 26.—Up early and started for the rookery. We separated after going inland for some distance, the mate and a sailor after a goat, and the rest of us for albatross. We

reached the first lot soon. They were a mile or more inland, on a smooth patch of ground. Some of the groups contained a dozen or more individuals. They were very tame, like the boobies, but some attacked

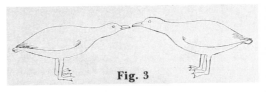

Fig. 3

us in a savage manner. We noticed a very curious and interesting habit which seemed to be a pastime of theirs, and resembled fencing as near as birds could imitate it— their beaks being the foils (Figs. 1 to 6). In every direction birds were fencing in pairs.

FIGURE 3.1. *The waved albatross.* The courting ritual of this elegant seabird is an intricate affair, captured in these sketches by Frederick Peabody Drowne in October 1897. 'They have a manner of fencing with their bills that is ludicrous and remarkable,' wrote Charles Harris, chief naturalist on the expedition. *Reproduced from Walter Rothschild and Ernst Hartert, Novitates Zoologicae 6, no. 2 (1899): 7–205; see Drowne's entry for 26 October 1897.*

Once the chick is strong enough to be left alone, it stands a good chance of surviving. Still, there's a lot of death amongst these hatchlings, with only around one in four breeding females rearing a chick to the point of fledging. Over the course of this chick-rearing phase, juvenile birds hatched in previous years start to return, and in November and December the colony reaches a crescendo of behaviour, with plenty of interactions between birds and courtship displays between existing pairs and wannabe breeders. This involves an impressive range of movements, from what looks like deferent head bobbing and circling, to jaw gaping and snapping, to a rapid rapier-like exchange of bill clacks, to a plaintive lowing thrown to the sky.

The Penguin and the Cormorant

In contrast to the waved albatross, with its long-distance movements, two other endemic seabirds do not fly: the Galápagos penguin and the flightless cormorant. The most suitable place for these species to feed and breed is in the productive west, between Isabela and Fernandina. Even then, they face similar pressures to the fur seals and sea lions. In the particularly harsh El Niño of 1982–1983, for instance, when fur seals and sea lions were so hard hit, only one in four penguins and cormorants survived.

For both the penguin and the cormorant, the unpredictable arena that is the Galápagos has had several consequences. The Galápagos penguin is famously small, with adults weighing in at just 2 kg, about half the size of their closest living relative, the Humboldt penguin from Chile and Peru. Only with its diminutive stature and relatively thin feathering can a bird most commonly associated with colder climes hang on so far north. The flightless cormorant, by contrast, seems to have grown in stature (it has more than twice the body mass of its closest living relative, the neotropic cormorant of Central and South America), possibly to improve on its capacity to dive. Whereas all other species of cormorant are pretty nifty swimmers, the flightless variety's heavier frame and more powerful feet mean that, for a given depth, it can stay down for longer. This shift away from flight could never have happened in a world full of predators. But in the Galápagos, where the only land-based threat to the flightless cormorant comes from the Galápagos hawk, the costly flight

FIGURE 3.2. *The Galápagos penguin.* This species is famously small, with adults weighing in at just 2 kg, about half the size of their closest living relative, the Humboldt penguin from Chile and Peru. *Reproduced from Osbert Slavin,* Transactions of the Zoological Society of London 9 (1877): 447–510.

apparatus—large flight muscles, huge wings—was essentially redundant. 'I believe that the nearly wingless condition of several birds, which now inhabit or have lately inhabited several oceanic islands, tenanted by no beast of prey, has been caused by disuse,' wrote Darwin.

As with much else in the Galápagos, the capricious conditions mean that the penguin and the cormorant must also be flexible when it comes to breeding. If the waters on the western edge of the archipelago are too hot (greater than 24°C to be precise), there simply won't be enough food for the penguins to breed. If the waters are cooler (22°C or lower), food will likely be sufficient, but breeding is still costly. Once they have put on enough weight, penguins will come ashore to moult, shedding old feathers and growing new ones. This takes between ten and fifteen days to accomplish, during which time they will not feed and may lose as much as a quarter of their body weight. Once a pair has located a suitable nest site, preferably one that's sheltered from the equatorial sun's egg-frying glare, the female will lay one egg (or a second if there's a lot of food to be had). If conditions take a turn for the worse, they will abandon first one of the chicks and then, if things get really bad, the other.

For cormorants, breeding is an even more intriguing affair. For most bird species, females typically invest more in reproduction than males, so are particularly choosy about whom they mate with (see blue-footed boobies and frigatebirds above). For the flightless cormorant, the roles are somewhat reversed. Although there is no way for the female to get around the cost of producing eggs (she'll typically lay three in a single reproductive event), she is far smaller than the male and, as a consequence, has pared down most of her other maternal duties to a minimum. It's the male that must build the nest, usually a fancy arrangement of seaweed harvested from the seabed. It's the male that brings most of the food to the newly hatched chicks. It's the male that's best able to keep predatory hawks away from the brood. And once the chicks have fledged (after a couple of months), it's the male that's left in the lurch when the female abandons him and her young offspring to go in search of a new partner. In a whimsical world, this is the female's way of increasing her chances of reproductive success. It might not seem particularly fair on the male. But it works.

If living off the oceans is a risky business, it doesn't come much easier on land. Let us leave the waves, the currents, the fish, the marine mammals and the seabirds and look to the rocky shoreline of the Galápagos. It is time to think about plants, where they came from, how they got here and how on earth they managed to survive.

Chapter 4. Plants

In 1965, a mat of seaweed broke away from Iceland's brutal southern shore and began to make for the wide, open Atlantic Ocean. But a couple of days later, this drifting vegetation hit land—new land—a small volcanic island that had started to emerge from the waves a couple of years earlier. Surtsey was still rife with volcanic activity; yet the seaweed found itself lodged in the volcanic debris on the island's north shore. In the carpet of organic matter, there was a seed of sea rocket, which settled into the virgin soil, germinated and began to grow. The appearance of this flowering plant, so soon after the formation of this island and whilst eruptions were still ongoing, is remarkable.

The botanists keen to witness the colonisation of a virgin volcano were certainly excited. Soon after the first eruption, they recorded the presence of bacteria, fungi and more algae, many presumably arriving on the wind as spores from Iceland itself some 40 km to the north. By the time the sea rocket rocked up, there was an ashy soil-like substrate to welcome it. Within a few years, other plants like sea sandwort and lyme grass had become established and started to flower. It makes sense that all of these early settlers should have been durable, able to survive the hazards of transport and the extraordinarily hostile landscape that awaited them.

So it was for the Galápagos too. Some of the first arrivals were likely windborne spores of bacteria, fungi, algae and lichens, carried across hundreds of kilometres of open water. These early settlers are particularly hardy, able to lodge on bare rock that most other species could not dream of calling home. They all contribute to the weathering of volcanic lava flows, bringing on the formation of soil-like substrates that give plants a realistic chance of survival.

As with the geology of these islands, our knowledge of Galápagos plants really starts with Darwin. 'Amongst other things, I collected every plant, which I could see in flower, & as it was the flowering season I hope my collection may be of some interest to you,' he wrote from Sydney, Australia, to his old mentor John Stevens Henslow in Cambridge. 'I shall be very curious to know whether the Flora belongs to America, or is peculiar.' With only limited botanical expertise, he needed help.

Henslow made a token effort, describing a couple of Darwin's *Beagle* cacti, one from Patagonia and the other from the Galápagos. He probably felt that this would satisfy his young friend. Yet Darwin had a far more ambitious goal: explaining where his Galápagos plants had come from. It was a project that, taken to its logical conclusion, would ultimately reveal the origin of species.

Darwin had collected sufficient specimens in the Galápagos to hit upon two key ideas. First, he realised that much of the Galápagos fauna had a distinctly South American flavour to it. Second, the species on one island seemed to differ slightly from those on the next. 'It never occurred to me, that the productions of islands only a few miles apart, and placed under the same physical conditions, would be dissimilar,' admitted Darwin in the first edition of his *Journal*, published in 1839.

His impressive collection of Galápagos plants proved to be a crucial test of this observation. So in 1838, he wrote again to Henslow to give him a prod. 'I do not want you to take any trouble in giving me names &c &c—all I want is to know whether in casting your eye over my plants, how many cases . . . there are of near species, of the same genus;—one species coming from one island, & the other from a second island.'

When Henslow again failed to produce the goods, Darwin got him to send the languishing *Beagle* specimens to up-and-coming botanist Joseph Dalton Hooker, who had recently returned from his own voyage of discovery to the Antarctic. Within days of receiving the Galápagos plants towards the end of 1843, Hooker wrote to Darwin in great excitement. There were indeed remarkable differences from one island to the next. In his initial analysis, Hooker judged that Darwin had handed him 217 different plant species, more than half of which were peculiar to the Galápagos. Of these so-called endemic species, the vast majority seemed to be confined to just one island. This fact, wrote Hooker, 'quite overturns all our preconceived notions of species radiating from a centre.'

Darwin, in turn, was ecstatic. 'I cannot tell you how delighted & astonished I am at the results of your examination; how wonderfully they support my assertion on the differences in the animals of the different islands.' So powerful was the case of the Galápagos plants that Darwin gave them a prominent place in the beefed-up second edition of his *Journal*, which appeared in 1845. 'Reviewing the facts here given, one is astonished at the amount of creative force, if such an expression may be used, displayed on these small, barren, and rocky islands.'

Seeding the Galápagos

This was all very well, but it still didn't tie up how the species had got there in the first place. The journey from South America to the Galápagos might be easy for us humans, catered for as we now are by more than forty flights a week. But for a plant it's a migration that presents a series of formidable challenges. Hooker mapped out several possible routes that bits of vegetation or seeds might have taken: by sea, by air, carried by birds or brought by humans.

The same strong current that carried the bishop of Panama to the Galápagos in 1535 will often bring miscellaneous vegetation from the continent. This much was noted more than two hundred years ago by James Colnett: 'On several parts of the shore, there was drift-wood, of a larger size, than any of the trees, that grow on the island: also bamboos and wild sugar canes, with a few small cocoa nuts at full growth, though not larger than a pigeon's egg,' he wrote in 1797. *Beagle* captain Robert FitzRoy also noted the way the driftwood was always to be found high and dry on the south-western shores of the islands.

Still, 1,000 km by sea? It's a heck of a way. Although Hooker observed that many of the Galápagos plants boasted large, tough seeds, something he figured would 'probably aid them in resisting for some time the effects of salt water', he remained sceptical that seeds could really withstand such a journey.

Darwin decided to find out and in 1855 began to play around with seeds and seawater at his home in the English countryside south of London. He kicked off with cress, radish, cabbage, lettuce, carrot, celery and onion, leaving seeds in saltwater for a week before planting them out. They all germinated, though some more convincingly than others. 'It is quite surprising that the Radishes shd have grown, for the salt-water was putrid to an extent, which I cd not have thought credible had I not smelt it myself,' he wrote. He went further, buying seeds of all manner of different species and gradually extending the length of time they were immersed in his seawater concoction. By the time he published *On the Origin of Species* in 1859, Darwin had exposed the seeds of eighty-seven different plant species to these hostile conditions. Incredibly, nearly all of them germinated after weeks stewing in brine; some still seemed in perfectly good working order after several months.

So seeds floating on the waves could easily account for the arrival of plants in the Galápagos. But in the course of his seed-salting experiments, Darwin hit upon another serious obstacle to this mode of travel. Most of the seeds he dunked sank immediately, and only a few bobbed back up to the surface. It made sense then that species with floatable seeds would stand a much better chance of taking a long-distance oceanic cruise from South America to the Galápagos.

Darwin hauled A. K. Johnston's *Physical Atlas* from his bookshelf, from which he worked out the average speed of several Atlantic currents. Based on all his observations, he concluded that 'the seeds of 14/100 plants belonging to one country might be floated across 924 miles of sea to another country'. The current that connects continental South America to the Galápagos can run considerably faster than

Darwin's Atlantic average, often around 100 km per day. So a back-of-the-envelope calculation suggests that a buoyant seed could reach the archipelago in just over a week.

More plants—even those with sinking seeds—might arrive by hitching a ride on a clod of earth attached to the roots of some big tree or carried in the carcass of a dead animal. To illustrate this possibility, Darwin fed a pigeon on seeds that would normally be 'killed by even a few days' immersion in sea-water'. He then sacrificed it and floated its body on salty water for a month. To his great surprise (and, one imagines, satisfaction), the delicate seeds, once dissected from the pigeon's crop and planted out, 'nearly all germinated'.

The Coastal Zone

Even if a floating seed does reach an island, it will face an immediate landscape that, for many, will be just too salty. It takes some nifty adaptations to make it in the coastal zone.

Plants can do this in a number of ways, some of them better suited to becoming established along the Galápagos shoreline. For a red mangrove, for instance, its roots are crucial. These drop down into the sea-water, acting like props to keep the trunk out of the water, and are so specialised that they can draw water from the sea without bringing too much salt on board. The leaves of black mangroves also contain salt glands capable of shifting salt from the inside to the outside of the plant. Mangroves and many other coastal species also have dark, waxy leaves, which reflect a lot of sunlight, preventing overheating and reducing unnecessary evaporation of precious water from the tissues.

Surviving is all well and good, but if you want to colonise, then you really have to reproduce. If, as was likely the case for most new arrivals in the Galápagos, you find yourself just one of a handful (or possibly the only one) of your species, the more self-sufficient your mode of reproduction the better. If you have a fussy reproductive set-up, like a flowering plant that relies on a special insect to carry pollen to a fellow member of your species, you could be in trouble. This helps explain why the vast majority of flowering plants in the Galápagos are capable of self-pollination and, with no need to attract insects, not particularly showy (white or yellow petals are the norm). Choose a Galápagos flower

at random. Take a peek at its reproductive parts. It's probably a her-
maphrodite, sporting both male and female apparatus. This is the case
for the red mangrove (named for its reddish bark rather than its yellow
flower). Unusually, it is viviparous too, which means that the seeds begin
to sprout before they are released from the parental tree. When these
so-called propagules finally drop, they can survive in salty water for more
than a year, obviously useful if they find themselves washed out to sea.

The Arid Zone

Alongside the seed-floating route to the Galápagos, Hooker imagined
other modes of transport too. Owing to 'the excessive minuteness' of the
spores of lower plants like mosses and ferns, he was in little doubt 'that
their diffusion by the winds is a never-ceasing though invisible operation', just as on the Icelandic island of Surtsey. Birds, he suggested, could
be important too.

Darwin agreed. 'Living birds can hardly fail to be highly effective
agents in the transportation of seeds,' he wrote in the *Origin*. Obvious as it might be, he still sought evidence, going out into his garden
in search of bird droppings. 'In the course of two months,' he wrote, 'I
picked up in my garden 12 kinds of seeds, out of the excrement of small
birds, and these seemed perfect, and some of them, which I tried, germinated.' In addition, the occasional seed might get stuck to a foot or a
beak. Although some Galápagos plants clearly came on ocean currents
and others on the wind, it's thought that most Galápagos plants probably
hitched a ride with birds.

If a settling seed is fortunate enough to land beyond the salty coastal
region, it could still be in for a shock. The lowest reaches of each Galápagos volcano are dominated by the arid zone, where water—or the lack of
it—is the perennial problem. It can be unbelievably hot too. On Santiago, Darwin stuck a thermometer into some brown soil, and the mercury
rocketed up to almost 60°C. It could have been yet still hotter, but Darwin's thermometer only went so far.

There are several ways to cope in this situation. As water escapes
through a plant's leaves, those species with fewer, smaller leaves and
fewer pores that open up to the atmosphere are going to be able to survive longer in an arid environment. This is why Darwin found the plants

beyond San Cristóbal's shoreline to be 'such wretched-looking little weeds', the landscape 'covered by stunted, sun-burnt brushwood'. From a distance, this appeared 'as leafless as our trees during winter'.

Herman Melville found the plant life more wretched still. 'On most of the isles where vegetation is found at all, it is more ungrateful than the blankness of Atacama,' he wrote in the first of his literary sketches on 'the Encantadas'. Whereas Darwin eagerly collected specimens to be pressed, Melville only saw 'tangled thickets of wiry bushes, without fruit and without a name, springing up among deep fissures of calcined rock and treacherously masking them'.

Many of these arid-dwelling species have interesting roots. Take the prickly pear cacti of the genus *Opuntia*, for example. These have two kinds: one superficial set of finer roots whose job it is to suck up every last drop of water in the aftermath of a downpour and a deeper 'tap' root, which gives it a stronger hold on the rocks and searches out deeper sources of water. The thorny *Acacia* bushes have also gone in for this approach, with a root network that grows during drought to cover a greater area of soil. Some of this spiky vegetation will be festooned with curtains of lichen, fine filaments that are able to trap moisture from the atmosphere. This then falls in steady drips to the earth around the tree.

Another approach is to specialise in water storage. This is something that cacti are famous for. The lava cactus is notable for its ability to establish on fresh lava flows, and specimens can be seen in Sullivan Bay on Santiago, in Punta Moreno on Isabela and around the rim of Genovesa. It's a stout cactus that comes in clusters. Its leaves have been so modified that we call them spines, a perfect way to cut down on water loss. This species is incredibly slow growing and doesn't often blossom. When it does, the window for pollination is just a matter of hours; then the petals shrivel, and the flower drops.

Compared with the lava cactus, the prickly pear opuntias are a much more common sight. It's well known that giant tortoises are long lived, and many visitors to the Galápagos often imagine that one of the tortoises Darwin saw back in 1835 might still be alive today, that they have seen one and the same animal. Although this seems vanishingly unlikely, it could certainly be the case for some of the biggest Galápagos opuntias. It's reckoned that these take around fifty years to reach maturity and start to produce flowers and fruits. They can then probably do so for a couple of hundred years.

In the Galápagos, there are six different species (with a couple further subdivided into distinct varieties). All of them have flattened oval branches (or pads as we tend to call them) that lobe off a central trunk and store the water. As with the smaller lava cactus, the leaves are no more than spines that stick out from the pad, considerably reducing loss of water.

Apart from these general similarities, there are plenty of differences between species. Darwin collected just one, a prickly pear opuntia specimen from Santiago, the one Henslow described back in 1837. But he would have been fascinated to have had a chance to look at the other forms that exist throughout the archipelago. The most obvious difference between species is their size. The Santiago species, at around 2m high with a trunk some 30 cm wide, turns out to be somewhere in the middle of the opuntia range. The northern islands of Darwin, Wolf and Genovesa are home to one of the smallest varieties, a low-lying form in which the trunk is largely obscured by a riot of pads. On Santa Fé, by contrast, there is a colossal species that can reach up to 12m high with a trunk that's over 1m wide.

It's often said that the different heights have come about as a result of the selective force imposed by the local herbivores. So on islands that have never had tortoises (like Darwin, Wolf and Genovesa, for instance), the local prickly pear opuntia is spread out near to the ground. On islands with tortoises, by contrast, the opuntias seem to have found a way to grow their succulent pads out of the reptiles' reach. Though possible, this explanation is probably a bit too simplistic. Although Santa Fé once had tortoises, would they really have driven the local opuntia to such a giant extreme?

The spines of the different *Opuntia* species also present something of a puzzle, one that would have intrigued Darwin. These differ markedly in number, length and strength. On Santiago, for instance, the fleshy pads are covered with clusters of between five and thirty-five spines that can measure up to 7.5 cm long. Henslow described these as 'strongly resembling hog's bristles.' On Genovesa, by contrast, the spines come in clusters of between seven and twenty-eight and are shorter and softer, more

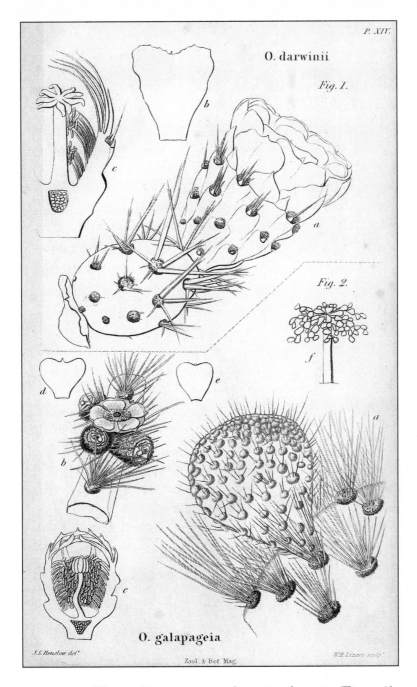

FIGURE 4.1. **The prickly pear cactus Opuntia galapageia (Figure 2).**
Charles Darwin collected this species from the lowlands of Santiago. It is
clearly different from another *Opuntia* he collected from mainland South
America (Figure 1). *Reproduced from John Stevens Henslow,* Magazine of
Zoology and Botany *(1837): 466–468.*

like sun-bleached hair. It's been suggested that these wavy spines have appeared on islands where there is no herbivorous threat; the plants just don't need the same defences they used to. Alternatively, soft spines may make it easier for birds like the cactus finch to come for pollen and nectar. On an island like Genovesa with few insect pollinators, such bird-mediated pollination could be particularly important.

If times get really tough, it's always possible to shut down completely. The prickly pear opuntias can do this by dropping pads, a strategy employed by another common arid-zone species, the palo santo tree. In the cool season from June to December, the palo santo is not in leaf, and its silvery lichen-clad bark shines out from the dark volcanic backdrop. When the hot season hits and rain begins to fall, its leaves begin to sprout, and the scattered stands become a much greener affair. The same goes for the deciduous guayabillo, pega-pega and matazarno that can be found further up, as the arid zone transitions into the moister highlands.

The Highlands

Given the challenges imposed by the coastal and arid zones, it seems extraordinary that plants should have made it any further. But with the aid of wind and birds (and indeed humans), plants can get to most places, and those islands that rise more than a few hundred metres above sea level boast some of the choicest plant habitats to be had.

During the cool season, between June and November, the *garúa* hovers at between 500m and 1,000m above sea level. This creates a lush moist zone, overrun with effulgent flora of an altogether different kind. Driving from Puerto Ayora up the road that cuts its way across Santa Cruz towards Baltra, this transition is quite clear. In and around the town, it's hot, it's dry and the vegetation is sparse. At the small town of Bellavista at around 200m above sea level, everything will have changed. There is no sign of lava any more. It is there, of course, but buried beneath a choking quilt of purest green. This is why Bellavista (and the smaller farming community of Santa Rosa) exists at all: the climate here is perfect for growing crops and grazing cattle. Travelling higher still, one finds a misty, luxuriant landscape carpeted with grass and sedge (see Appendix C, Figure 5).

When Darwin experienced this transition, it struck him too. Landing at Black Beach on Floreana (where the small town of Puerto Velasco Ibarra now stands), he and FitzRoy were met by the vice governor of the islands, Nicholas Lawson, who then walked them around 8 km inland to the small penal colony he was supervising on behalf of the Ecuadorian government. 'The wood gradually becomes greener during the ascent,' Darwin wrote in his diary. 'Passing round the side of the highest hill; the body is cooled by the fine Southerly trade wind & the eye refreshed by a plain green as England in the Spring time.' After the barren landscapes of Peru and Chile, he was thrilled 'to find *black* mud & on the trees to see mosses, ferns & Lichens & Parasitical plants adhaering.'

He was in *Scalesia* forest, a habitat named after the giant daisy tree *Scalesia pedunculata*. This is a species that would normally dominate the vegetation at this altitude, though it—and the wider habitat—has been severely affected by clearance for agriculture and pressures from introduced species. On Santa Cruz, the visitor site known as Los Gemelos boasts a relatively intact patch of highland forest, the upstanding S. *pedunculata* standing guard above the turmoil of vegetation tumbling into the depths of these collapsed twin craters. The trees grow incredibly quickly, dwarfing a two-story house within a matter of years as they shoot up in search of the sun. Their leaves cluster at the skyward tips of the branches to form a kind of canopy, and just beneath is a crowded orgy of mosses, liverworts, orchids, and passion flower, all using the trunk and branches as a kind of ladder. There are other trees here, including Galápagos pisonia, the Galápagos guava and cat's claw, though they struggle against the dominant position that S. *pedunculata* has assumed.

At around five years old and about 8m tall, the giant daisy tree will begin to flower. Tiny white petals emerge on top of the canopy. The seeds will fall to the forest floor, but in this shady environment they will not germinate until their parent plants have come to the end of their fast, short lives at about age fifteen. Then, as the canopy starts to die back and sunlight streams to the floor, the seedlings and saplings emerge, all racing together to replace their parents as if in some kind of shared adventure. Between each unfolding canopy, the trees maintain daylight, a gap of around 30 cm. If I may indulge in a touch of anthropomorphism,

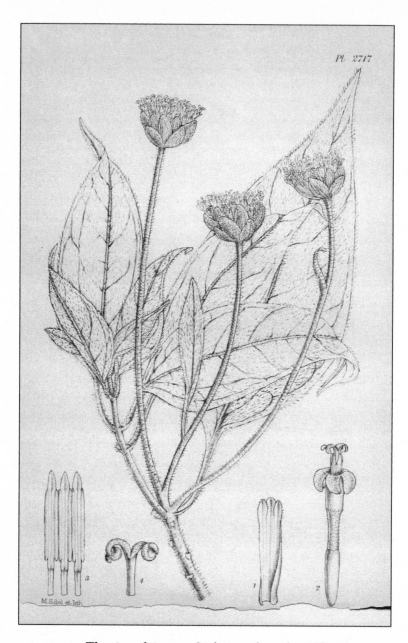

FIGURE 4.2. ***The giant daisy tree* Scalesia pedunculata.** This is just one of at least fifteen closely related species found only in the Galápagos, but it is certainly the most spectacular, a slender tree that grows rapidly to a height of some 20m. In the past, this species would have dominated much of the highlands—'a very handsome species', according to botanist Joseph Hooker. *Reproduced from Joseph Hooker,* Icones Plantarum 28 *(1905): pl. 2717.*

it is as if each specimen is acutely aware of its neighbour's sensitivities, the forest a paragon of mutual respect.

Scalesia pedunculata is not just a pretty tree; it is also part of a fascinating evolutionary story. For it's not the only *Scalesia* in town. In fact, the genus comprises at least fifteen different species (some of which are further divided into subspecies), all of them endemic to the Galápagos and found from the arid zone to the highlands. In spite of considerable differences in outward appearance, its thought that all these species are probably descended from a single daisy-like ancestor that managed to establish in the arid zone. From this unpropitious origin, subsequent generations spread higher and higher, adapting to the increasing humidity along the way and dispersing to other islands. Small, low-lying islands tend to support one shrubby *Scalesia* species. Larger, elevated islands have more. On Santa Cruz, for instance, there are at least six different species.

Beyond the *Scalesia* forest, on established islands like San Cristóbal and Santa Cruz that rise above 400m, there are more rarefied habitats still. There are no *Scalesia* here. They give way to lower-lying, bushy thickets of *Miconia* (the nesting site for the critically endangered Galápagos petrel), then, even higher, to the pampa zone, a boggy moor-like expanse of grass, sedge, moss and fern.

All this vegetation, from the salty coast, through the desert-like arid zone and the humid highlands, to the chilly, thinned-out pampa, provides the backdrop for the arrival of land-dwelling creatures from tiny insects to hulking reptiles to humans. It is to them that we now turn.

Chapter 5. Invertebrates

Most visitors to the Galápagos will overlook its smallest inhabitants in favour of the larger, more charismatic species, like the blue-footed booby. But Charles Darwin was not your typical visitor. He'd had a love of beetles from an early age and described his collecting during his Cambridge University days as the activity that gave him the most pleasure. One day, out in the countryside, he'd grabbed hold of two rare beetles from beneath a chunk of bark, holding one in each hand. On spotting a third species, he wanted to free up a hand and raised his right fist to his mouth and let the beetle crawl in. 'Alas! it ejected some intensely acrid fluid, which burnt my tongue,' he wrote, forcing him to spit it out and return with just the beetle in his left hand.

Insects

When he first set foot in the Galápagos, Darwin was most interested in the rocks, but he kept a look out for beetles and other invertebrates too.

When in the highlands, he repeatedly scoured the bushes for insects in all kinds of weather but only came across a few species, moaning that with the exception of the barren forests he'd explored in Tierra del Fuego, he had 'never collected in so poor a country'. Of his beloved beetles, for instance, he only found twenty-nine different species. This may sound like a good return on five weeks' work, but he would certainly have expected more. Indeed, there is a pretty tight relationship between the area of an island or archipelago and the size of its beetle fauna. The bigger the island, the more beetles. But the Galápagos sits well below this line, boasting the number of beetle species you'd expect of an archipelago half the size. This is probably down to a combination of factors: the relatively recent origin of the islands and the lack of water.

If beetles have struggled to settle in the Galápagos, so too have other insects that would normally thrive at this latitude. Take butterflies for instance. Over the water in mainland Ecuador, there are well over 2,000 different species. In the Galápagos, there are just ten. The paucity of butterflies looks even stranger when you consider that there are more than three hundred species of moth—except, that is, if moths, being principally active at night when the world is cooler, just find it easier to get established. It's certainly a possibility.

The intensity of the equatorial sun may also explain why there's just one species of bee in the islands. If others have attempted to colonise, they appear to have failed in their efforts. That the Galápagos carpenter bee managed to survive could well be down to its eclectic tastes. Whereas many other bees show a strong preference for a particular flower, the Galápagos carpenter bee seems happy to feed from an impressive array of flowering plants. In the Galápagos, it does not pay to be too fussy. As it forages for nectar and pollen in the coastal and arid zones, this species may play a crucial role in the pollination of many plants. Once it's had its fill of nectar, an all-black female will carry pollen back to her larva, a single individual nestled in a hole bored into the branch of a palo santo (or other suitable) tree. American naturalist William Beebe remarked on the strange absence of insect voices. 'Even when an occasional large, black bee appeared, it flew and drew nectar from the flowers silently, with muffled wings,' he wrote in *Galápagos: World's End*, the best-selling account of his first visit to the islands in 1923. 'The trade winds made no

FIGURE 5.1. *Butterflies and moths collected by the California Academy of Sciences in 1905 and 1906.* The large moth at the top (No. 7), *Manduca sexta leucoptera*, was collected as a caterpillar in Wreck Bay on San Cristóbal. *Reproduced from Francis Williams*, Proceedings of the California Academy of Sciences, Fourth Series 1 *(1912): 289–322.*

sound among the thin foliage, as if they had blown so long and so regularly that no sough was left in this part of earth.'

In contrast to the Galápagos plants, whose most common mode of transport to the archipelago was hitching a ride with birds, most insects—certainly the butterflies, the moths and the bee—are thought to have taken the aerial route. Even spiders, if small enough, can use strands of silk to parachute through the air. Darwin had noticed this earlier in the voyage when the *Beagle* was almost 100 km from land. He watched in amazement as thousands of tiny, dusky red spiders came floating on board, their silk threads trapping them in the rigging. This and observations of other spiders doing much the same led Darwin to the conclusion that 'the habit of sailing through the air' was 'characteristic of this tribe'. In 1992, in the midst of an El Niño, one entomologist set up nets from boats whilst travelling between islands in the Galápagos. The stormy conditions blew more than 18,000 specimens of insects and spiders into his hands.

Larger insects not blessed with the power of flight or a silky parachute, like many beetles, bugs, mantids, weevils and crickets, probably came by sea, rafting on mats of vegetation or possibly even floating on the surface itself. Other invertebrates, like centipedes, mites and ticks, were probably carried by birds. Although most of this creepy crawly fauna is small, this is not always the case. When it comes to size, one of the Galápagos centipedes deserves a special mention. It goes by the name of *Scolopendra galapagoensis* and can reach up to 30 cm long. As with other centipedes, its front legs are modified into a pair of devilishly sharp pincers that it uses to inject a venomous cocktail into its prey of insects, lizards and sometimes birds. A couple of ornithologists, camping out on isolated Wolf Island in the early 1980s, found this 'foot-long centipede' a particular nuisance. 'Slowly crawling about at night, when frightened they swiftly dash for cover and are usually found in the morning curled up in our cooking utensils or food cans.' There are a couple of scorpions in the Galápagos too. Their stings, though painful, are usually not fatal to humans.

There are, of course, several entire branches missing from the Galápagos tree of life and the part of the canopy in which the invertebrates reside is no exception. There are, for instance, no mayfly, stonefly, caddisfly, or alderfly. It would be rather surprising if there were, dependent as all these insect families are on freshwater for the development of their larvae. As Bishop de Berlanga and his men found, this is just not something that you find much of in the Galápagos.

For those land-based invertebrates that did make it, the special set-up in the Galápagos has resulted in speciation, and hence the origin of new species, on a particularly impressive scale. Some 735 of 1,555 native insects, for instance, are endemic, found nowhere else in the world. The ratio is higher for spiders, with fifty-five out of around eighty species unique to the Galápagos. The Galápagos flightless weevils are more special still, with ten of thirteen recognized species originating in the islands. In fact, looking at terrestrial invertebrates as a whole, over half of all native species are endemic, made in the Galápagos.

Land Snails

When it comes to speciation in the Galápagos, though, one invertebrate family tops them all: the bulimulid land snails. All of them are pretty small (less than 3 cm) and difficult to see (often hiding out beneath boulders of lava), and they might not be the sort of thing that really excites you. But if, as has been suggested, a single ancestral colonist has given rise to around seventy different species to be found from the arid zone to the highlands, the Galápagos bulimulids might offer one of the finest examples of adaptive radiation anywhere in the world.

Darwin was interested in snails, boxing sixteen different species (including some bulimulids), and he got to thinking about how they might have reached remote islands like the Galápagos. 'It occurred to me that land-shells, when hybernating and having a membranous diaphragm over the mouth of the shell, might be floated in chinks of drifted timber across moderately wide arms of the sea,' he wrote. Bulimulid snails, like other land snails, can indeed seal up their shells to prevent desiccation, which would have been helpful for rafting out across the Pacific. But as anyone who has tortured slugs knows, salt is something that snails and their allied molluscs do not relish. Perhaps there was another way, 'some unknown, but highly efficient means for their transportal,' mused Darwin. 'Would the just-hatched young occasionally crawl on and adhere to the feet of birds roosting on the ground, and thus get transported?' he wondered.

It was a pretty good idea, and Darwin hit upon another of his delightful, 'Heath Robinson' ways to test it. Working from home at Down House in Kent post-*Beagle*, he took one duck's foot (dismembered) and dangled it in an aquarium in which some freshwater snails had recently hatched. Within a matter of hours, dozens of minute snails were tenaciously stuck to the foot. No amount of waving it about could dislodge them, and a few survived for more than twelve hours out of water. In this time, Darwin figured, 'a duck or heron might fly at least six or seven hundred miles, and would be sure to alight on a pool or rivulet, if blown across sea to an oceanic island or to any other distant point.'

We now know that land snails can do this too, worming their way in amongst the feathers of birds. Incredibly, snails might even be able to survive a journey through an avian intestine. In a recent study, researchers fed land snails to a couple of different bird species and found around one in six of them made it out the other end intact. One snail even emerged and promptly gave birth. This kind of hitchhiking makes more sense than rafting on the ocean. A snail that washed up on a shore would have struggled to make it across the salty, barren landscape to the more suitable habitat beyond. It is notable that bulimulid snails are completely absent from the coastal zone, and the case of Fernandina, where there's a vast stretch of impassable lava between the shore and the safety offered by vegetation, is particularly telling. The snails that live on the verdant rim of its crater simply couldn't have got there by walking. Conclusion: they must have been airlifted.

Divergence

So how could one colonising species have diverged into all the different bulimulids we see today? The two key ingredients for this to occur are isolation and time. With enough of these, two populations will diverge to become two species, such that if they find themselves reunited they no longer recognise each other as kin. For a small, water-dependent creature that moves at a snail's pace, the landscape is full of obstacles: a freshly solidified lava flow, a patch of dry scrub, a deep crevice. With so many opportunities for isolation and several million years to play with, one species can easily become many.

In order to understand the mechanics of this divergence, it's necessary to know a little genetics. The production of sperm and eggs is achieved through cell division. This demands the duplication, division and repackaging of the DNA contained in the nucleus of a cell. During all this DNA rearrangement, the odd mistake can creep in. So after many generations in isolation, the genetic makeup of one population is likely to have 'drifted' in a different direction to that of another. But how could such ostensibly meaningless changes in a genetic sequence give rise to differences in appearance and behaviour that have so much meaning? How, for instance, could this process of random genetic drift possibly account for the fact that those bulimulid snails living at altitude have a classical rounded look to their shell, whereas those at lower elevations have more slender, pointy shells? The short answer is that it can't.

We need instead to consider a second mechanism that can lead two isolated populations in different directions, a phenomenon far more powerful than genetic drift. Snails trying to make it at different altitudes will face rather different pressures. At the highest altitudes, where there's plenty of moisture, the default, rounded shell works just fine. At lower altitudes, however, this is not the case. As conditions become more arid, a well-rounded snail with its relatively large shell aperture will find life increasingly hard, with a very real risk of dehydration and death. A snail with a slightly pointier body whorl (a feature that is under genetic control) will tend to have a tighter opening to its shell, a characteristic that will put it at a distinct advantage in such droughty conditions. Those snails that manage to survive, therefore, will have a genetic makeup that is far from random (as it is with genetic drift). The hardy survivors will have been 'naturally selected', not by some forethinking, supernatural selector but by virtue of the simple fact that they didn't die. At some point, which appears to be around 40m above sea level, the conditions become so harsh that few snails—however pointy they might be—seem able to survive.

It would be rather nice to test this explanation with an experiment, moving fat snails down to the arid zone to see how they fared. Prediction: the fat snails would struggle and die. Unfortunately, the Galápagos bulimulids have undergone such alarming declines in recent decades that this life-or-death experiment would simply not be possible. More than fifty species are now listed on the World Conservation Union's Red List of Threatened Species, many of them considered critically endangered and several of them probably extinct. We cannot know for sure the reasons, but species that occupy and have become tied to tiny ranges are especially vulnerable. If there's rapid change of any kind, like the

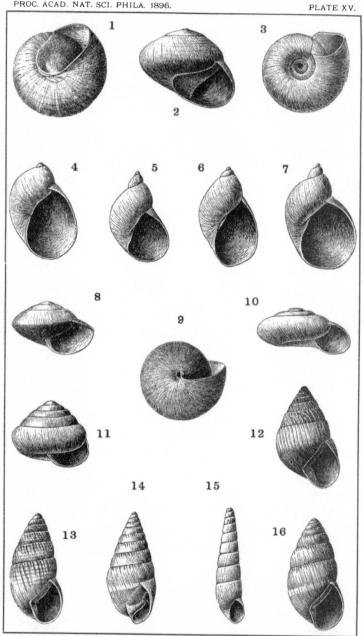

DALL. INSULAR LAND SHELL FAUNAS.

transformation of highland habitat into agricultural land or the introduction of invasive predators like rats, the disturbance can be devastating.

As brilliant as this family of snails is at illustrating the principle of evolution by natural selection, the land birds of the Galápagos have made the most significant contribution to the public understanding of Darwinian evolution. It is to them that we now turn.

FIGURE 5.2. **Left,** *land snails.* There is an incredible diversity of land snails in the Galápagos, almost all of them endemic to the islands. The Galápagos bulimulid family (to which specimens 12 to 16 belong) is one of the most remarkable examples of adaptive radiation. *Reproduced from William Healy Dall*, Proceedings of the Academy of Natural Sciences of Philadelphia 48 (1896): 395–460.

Chapter 6. Land Birds

If there's one group of Galápagos animals that's become inextricably associated with Charles Darwin, it's the finches. This is rather odd, because Darwin made very little mention of this group of rather ordinary looking little land birds. This raises several interesting questions: What did Darwin make of the Galápagos finches when in the islands? Why didn't he make more of them in his subsequent writing? When, exactly, did people start to refer to them as Darwin's finches? And why? Over the past several decades, historians have had plenty of fun finding the answers.

An Inexplicable Confusion

During his visit, Darwin found himself unable to make head or tail of the finches. 'Amongst the species of this family there reigns (to me) an inexplicable confusion,' he confessed in his *Ornithological Notes*, written around nine months after he left the Galápagos. For anyone who's

been to the Galápagos and hoped but failed to distinguish one finch species from the next, it should be of some consolation that this was Darwin's experience too. For a start, he didn't appreciate that they were all
finches, judging the cactus finch to belong to the family that contains
New World blackbirds and the warbler finch to be a kind of wren. Of
those he recognised as finch-like birds, he found them rather samey, a
mass of feathers ranging from light brown for the females to dark brown
or black for the males, with no obvious markings to aid identification.
Their behaviour also seemed rather unremarkable: 'There is no possibility of distinguishing the species by their habits, as they are all similar, &
they feed together . . . in large irregular flocks.'

But though identifying the Galápagos finches with any certainty is a
feat best left to the serious ornithologist, the casual visitor will have no
trouble seeing that the beaks of these birds come in an impressive range
of shapes and sizes, from the very neat (as in the case of the warbler
finch) to the frankly huge (like that of the large ground finch). Darwin
saw this too, noting 'a gradation in form of the bill'.

With his mind focused on geology, however, Darwin set his finch
specimens aside to be described at a later date by someone who really
knew his birds. That person turned out to be ornithologist and taxider-
mist John Gould, who judged that Darwin's thirty-one finch specimens
belonged to thirteen different species. This came as something of a surprise. Darwin, quite rightly, began to wonder how so many seemingly
similar species could live alongside each other. He flipped back through
his notes, wondering if he could figure out on which island he'd shot
each of his specimens. He couldn't. He got in touch with Robert Fitz-
Roy and a couple of other *Beagle* hands who had assembled their own
private finch collections and whose recollection of where each specimen had come from was more reliable than his own. In the end, however, the most he ever made of the Galápagos finches (at least in public)
was in the beefed-up second edition of his *Journal of Researches*, where
he drew attention to the rather impressive variation in the shapes and
sizes of their beaks. 'Seeing this gradation and diversity of structure in
one small, intimately related group of birds, one might really fancy that
from an original paucity of birds in this archipelago, one species had
been taken and modified for different ends,' he wrote. To illustrate his

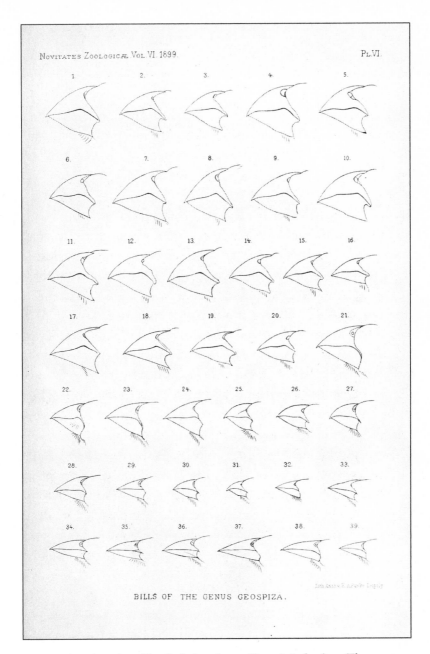

Novitates Zoologicæ Vol. VI. 1899.   Pl. VI.

BILLS OF THE GENUS GEOSPIZA.

FIGURE 6.1. *A series of beaks belonging to Darwin's finches.* These unassuming birds have become ambassadors for evolutionary biology. This plate shows off the 'gradation in form of the bill' that Darwin observed, from the large ground finch (top left) to the sharp-beaked ground finch (bottom right). *Reproduced from Walter Rothschild and Ernst Hartert,* Novitates Zoologicae 6 (1899): 7–205.

point, he included a rather nice woodcut of four finch heads, including those of the dinkily billed warbler finch at one extreme and the chunkily beaked large ground finch at the other. But as he'd only recorded the island of origin for a couple of his specimens and had not detected much difference in their behaviour, there was not much more he could say. He had little option but to leave the finches out of On the Origin of Species altogether.

The Mockingbirds

Thankfully, another group of Galápagos birds was much easier to make sense of than the finches: the Galápagos mockingbirds. 'I fortunately happened to observe, that the specimens which I collected in the two first islands we visited, differed from each other, and this made me pay particular attention to their collection,' he wrote in Zoology of the Voyage of H.M.S. Beagle. So when Darwin subsequently stepped ashore on Isabela and then on Santiago, he made a special point of collecting a mockingbird from each. In total, Darwin sailed away from the Galápagos with just four birds, one from each of the islands he had visited, but he also had a chance to study other mockingbird specimens collected by FitzRoy and others. He was quick to note the similarity between the Galápagos mockingbirds and those on the South American mainland and that 'each variety is constant in its own Island'. This observation led to Darwin's first explicit suggestion that species might not be fixed in their nature but might in fact change. If he really was able to demonstrate that each island had just one type of mockingbird that differed from those of a neighbouring island, it would, he felt, 'undermine the stability of Species'.

And so it did.

There are four species of mockingbird in the Galápagos, one occurring on San Cristóbal, another on Española, a third on Floreana and the most ubiquitous found on most of the other islands in the archipelago. There's not much to separate them visually. The Floreana mockingbird is the most distinct, with a notable pale patch behind its eye and three white bands on its wing covets, features that may have piqued Darwin's interest all those years ago. Unfortunately, the chances of seeing this particular species today are virtually nil. By the end of the nineteenth

FIGURE 6.2. *The Floreana mockingbird.* With a notable pale patch behind its eye and three white bands on its wing covets, the Floreana mockingbird clearly differs from mockingbirds on other islands. This kind of variation from one island to the next, Darwin felt, would 'undermine the stability of Species'. *Reproduced from John Gould, 'Part 3. Birds,' in* The Zoology of the Voyage of H.M.S. Beagle, *ed. Charles Darwin (London: Smith, Elder and Co., 1838–1843), 62.*

century, it had disappeared from Floreana altogether, probably as a result of depredation by introduced rats, cats and dogs and destruction of its habitat by goats. It survives still in two isolated populations on the tiny satellite islets of Champion and Gardner-by-Floreana. These are highly protected sites and off limits to tourists, but visitors to any of the other major islands are likely to be welcomed by its mockingbirds. In contrast to most other species in the Galápagos, which will pay humans little attention, these natural scavengers, which feed on everything from seeds to sea lion placenta, are always on the lookout for a stray morsel or drop of water.

So suggestive of evolution were Darwin's Galápagos mockingbirds that in the *Origin* he painted a picture of how a few birds from mainland South America might have reached the archipelago, their descendants gradually populating other islands and adapting to the slightly different conditions found on each. From there he broadened out his argument: 'We see this on every mountain, in every lake and marsh,' he announced with a characteristic flourish.

Darwin's Finches

In spite of the role that the Galápagos mockingbirds played in Darwin's thinking, these mischievous little birds have been largely eclipsed by the Galápagos finches, at least in popular culture. Yet it took more than a century before biologists really picked up on Darwin's hint and began to study this in-your-face feature—the beak—in earnest. In his 1947 book *Darwin's Finches*, British ornithologist David Lack was the first to make a really good case for a relationship between beak morphology and feeding habits.

In preparation for his visit to the Galápagos in late 1938, Lack had read a couple of travel books but still found himself taken aback by 'the inglorious panorama'. Dropping anchor just off San Cristóbal, he surveyed the small settlement that has since grown to become Puerto Baquerizo Moreno, the administrative capital of the archipelago. 'Behind a dilapidated pier and ramshackle huts stretched miles of dreary, grey-ish brown thornbush, in most parts dense, but sparser where there had been a more recent lava flow, and the ground still resembled a slag heap. The land rose gradually, with no exciting features, to a sordid cultivated region, beyond which, partly concealed in cloud, were green downs, the only refreshing spot in the scene.'

In the months to come, Lack found the Galápagos more depressing still. 'The biological peculiarities are offset by an enervating climate, monotonous scenery, dense thorn scrub, cactus spines, loose sharp lava, food deficiencies, water shortage, black rats, fleas, jiggers, ants, mosquitoes, scorpions, Ecuadorean Indians of doubtful honesty, dejected, disillusioned European settlers.' The Galápagos finches were not much to write home about either: 'dull to look at, not only in their orderly ranks in museum trays, but also when they hop about the ground or perch in the trees of the Galápagos, making dull unmusical noises'. He was, however, excited by 'the variety of their beaks and the number of their species'.

The most common species, which can be seen in the arid zones of most islands, are the small, medium and large ground finches. These little birds cannot be reliably distinguished on the basis of their size or plumage, but Lack found that their classically finchy beaks suited to cracking open seeds came in three clearly different sizes. The cactus finch also feeds on seeds, though its slightly longer bill allows it to dig into the fleshy pads of *Opuntia* cactus for water and probe its yellow flowers for nectar. The relatively large vegetarian tree finch uses its downward curving beak to feed on leaves, buds, blossoms and fruits. Then there are the small, medium, and large tree finches, concentrated in the fertile highlands where they flit amongst the branches rather like great tits, using their tough, parrot-shaped beaks to hunt down insects. The woodpecker finch operates in a similar habitat and is similarly insectivorous, though its longer, sharper beak allows it to drill into trees for more inaccessible grubs. Although it cannot reach inside the hole with its tongue to extract a tasty morsel (as would a true woodpecker), its cunning workaround is to pick up a cactus spine or twig in its bill and use it to achieve the same end. The diminutive warbler finch's neat, tweezer-like beak is just perfect for plucking small insects from low-lying vegetation and sometimes even from the air.

There are even more unusual feeding habits. In 1964, ornithologists following in Lack's footsteps were the first to observe the rather gruesome behaviour of the sharp-beaked ground finch on Wolf Island. This species occurs throughout the archipelago, but on this north-westerly island, it also goes by another name: the vampire finch. It has developed a taste for blood. With their devilishly sharp bills, these small birds hop onto the back of a booby (of the red-footed or Nazca variety) and peck away at the base of the seabird's feathers until the blood begins to flow. Then they will drink. The vampire finches will also occasionally use their bill as a lever to roll eggs out of a nest. Some even turn their beak into a kind of spring, burying it in the sand or beneath a rock and then launching backwards to give a booby egg a good kicking. Once the shell is cracked, there is something of a feeding frenzy, with several finches squabbling over the embryonic spoils.

Based on the analysis of DNA, the family of Galápagos finches is yet young, with a single species of proto-finch reaching the Galápagos somewhere between 2 and 3 million years ago. We don't know what this species looked like for it no longer exists, but its descendants do in the shape of the fourteen species we recognise today. The lineage that led to the warbler finch was the first to appear somewhere between 1.5 and 2 million years ago, with other species appearing soon after. Of the main islands, Española has the fewest finches, with just three species. At the other extreme, both Santiago and Isabela have ten different species of finch.

This pattern is in stark contrast to that of the mockingbirds, where each island has just one species. It is just a guess, but one reason there are so many more finches than mockingbirds might be because the finches have a relatively fixed and limited song. Closely related finches notice and act on differences in repertoire, preferring to mate with individuals with a more familiar song; hence two different forms may emerge on the same island. As mockingbirds have a much more varied and flexible song and continually learn new snippets throughout life, two birds that live on the same island but sound completely different will still mate like they belong to the same species.

The Grants

Since Lack's pioneering work, dozens of researchers have been drawn to the Galápagos finches in an attempt to make sense of this 'inexplicable confusion'. The result is a remarkable body of work that has given us a unique insight into the process of evolution, a clear illustration of natural selection, adaptation, divergence and the origin of new species. The most significant contribution has come from two British ornithologists, Peter and Rosemary Grant, who began their study of the Galápagos finches a month before Lack's death in March 1973. 'In a sense, we feel we are the bearers of a torch he passed on,' they wrote in *How and Why Species Multiply* in 2007.

Flying into Baltra and waiting in the queue for immigration, one can see to the west the craterous outline of Daphne Major, an island about the size of thirty football pitches where the Grants set up their operational base over forty years ago. Since then, they have been back—year in year out—typically spending six out of every twelve months furthering our understanding of these birds. 'This is one of the most intensive and valuable animal studies ever conducted in the wild,' wrote Jonathan Weiner in his Pulitzer Prize–winning classic *The Beak of the Finch*. 'It is the best and most detailed demonstration to date of the power of Darwin's process.' This statement holds as true today as it did when Weiner was writing some twenty years ago.

In the first half of the 1970s, Daphne Major received plenty of rain, resulting in abundant vegetation and plenty of seeds. Concentrating on the medium ground finch, the Grants caught and banded more than 1,500 of these birds, but a drought in 1977 caused a dramatic crash in the population, with fewer than 2 in 10 birds surviving. What was really interesting, however, was that the surviving finches had bigger beaks than those that didn't make it. The reason was a dearth of small, soft seeds of which the medium ground finches of Daphne would normally have had their fill. Only those individuals with the biggest bills were able to crack into the still abundant larger seeds. In the space of just a few months, the grim reaper's uneven swipe through the population had caused the average depth of the medium ground finch's beak to jump. Before the drought the average beak depth of the medium ground finch was less than 9.5 mm; afterwards it was more than 10 mm. Half a millimetre might not sound like much, but in relation to the size of these beaks, where the differences from one species to the next can be a matter of just 1 or 2 mm, it's huge. As beak size and shape are under genetic control, the big-billed survivors passed this trait on to the next generation.

A few years later, the tables were turned with the strong El Niño of 1982–1983 that so disrupted the lives of marine creatures throughout the archipelago. The enormous amounts of rain resulted in rampant vegetation and a resurgence of small-seeded plants. Under these conditions, the medium ground finches with the pointiest beaks hoovered up, a fact reflected in the next generation of pointier billed finches.

The weather was not the only influence on the medium ground finches of Daphne Major. In the wake of the 1982–1983 El Niño, a few large ground finches reached the island for the first time. So when the next serious drought occurred in 2003–2004, the medium ground finches with the biggest beaks were unable to muscle in on the supply of larger seeds, and most of them died. The presence of two competing species effectively carried the average beak size in different directions, with that of the medium ground finch getting smaller and that of the large ground finch getting slightly larger still.

So factors like this—food abundance and competition with other species—can have a dramatic effect upon the average beak size of a population, causing it to change from year to year in subtle and sometimes exaggerated ways. For those still in doubt about what's driving this change, it's hard to argue with the simple fact that today's medium ground finches on Daphne Major are smaller and have pointier bills than they did forty years ago. Then reflect that this kind of change, which even humans can detect with their relatively crude measuring tools, is by no means unique to this one species on this one island. Indeed, the average beak in every finch population of every species on every island that researchers have taken the time to look at has been in a state of flux.

So we know that the beaks of finches can change in what is effectively an evolutionary instant. But how does this give rise to new species? The most obvious mechanism is that in a relatively short period, two populations of the same species living on different islands (or maybe even on different parts of the same island) can end up looking rather different. When individuals from the two populations get together again, as they will inevitably do from time to time, they treat each other with something like suspicion, preferring to mate with birds that look and sound more like those they grew up with.

A neat experiment conducted in the early 1980s by one of the Grants' many students simulated this kind of chance reunion. Using museum specimens stuffed in attractive poses and perched at either end of a stick, the researcher gave male finches a choice. In each trial, one of these solicitous dummies was a female from the same population as the territorial male, and the other was a very similar female from another island. It's hard to imagine the males were not a little excited at the sudden appearance of two attractive and apparently available females, but once they'd got over their surprise, they showed a clear preference for the local female.

More important even than looks is how a finch sounds, and finches learn their relatively simple vocal repertoire from their parents early in life. By playing back sound recordings from local and more distant populations, it's been possible to demonstrate that different finch species—even if remarkably similar to look at—can and do distinguish each other on the basis of their song. Only with this kind of discrimination can two populations become two species.

Yet such barriers to reproduction are not insurmountable, and one species of Darwin's finch will occasionally mate with another. These unorthodox couplings, say between a medium ground finch and a cactus finch, do pretty well at first, most nests producing roughly the same number of fledglings as usual, though these hybrid youngsters struggle to make it into adulthood. On Daphne Major between 1976 and 1982, for instance, no hybrid offspring survived long enough to breed themselves. But with the arrival of the powerful 1982–1983 El Niño, which transformed the habitat, hybrids came into their own and did as well as, if not better than, either of their parental species. When this hybrid cohort came to breed, they had no trouble finding mates and produced good numbers of eggs, which hatched into healthy chicks and then fledglings.

This hybridization could play a very important role in the radiation of the Galápagos finches. When a hybrid manages to breed, backcrossing with one or other of its parental lineages, it injects some really valuable genetic combinations into the mix. This is likely to strengthen the genetic architecture of subsequent generations. It could even send a lineage in an entirely novel evolutionary direction.

From this brief survey of the Galápagos finches, it should be pretty obvious that the notion of a species is rather artificial. Darwin was well aware of the many different ways that his naturalist friends chose to define species: 'In some, resemblance seems to go for nothing . . . in some, descent is the key—in some, sterility an unfailing test, with others it is not worth a farthing,' he wrote to his botanist chum Joseph Hooker. 'It all comes, I believe, from trying to define the indefinable.' In the case of the finches, the number of species that we settle on is a balancing act, an effort to acknowledge both similarity and difference simultaneously. Being confused by Darwin's finches is not a failing but a strength, an honest acknowledgement of the beakish continuum that befuddled Darwin himself.

Hawk and Dove

Apart from mockingbirds and finches, there are plenty of other interesting land-dwelling birds in the Galápagos. The Galápagos hawk, the only raptor and the top predator on the islands, is particularly so. Darwin found it remarkable for its vulturous habits, as reported in this rather gruesome passage: 'When a tortoise is killed even in the midst of the woods, these birds immediately congregate in great numbers, and remain either seated on the ground, or on the branches of the stunted trees, patiently waiting to devour the intestines, and to pick the carapace clean, after the meat has been cut away,' he wrote.

In fact, the resemblance to vultures is merely superficial. John Gould judged the Galápagos raptor to be similar to actively hunting hawks from the Americas. Recent genetic work shows that this is right and that the closest living relative of the Galápagos hawk is Swainson's hawk. So great, in fact, is the similarity between these species that the hawks of the Galápagos are among the most recent arrivals in the archipelago, a few birds blown off their long-distance migratory path between North and South America within the last few hundred thousand years.

In that time, the descendants of these first hawkish settlers have styled themselves on vultures, with a more passive, wait-and-see approach to finding their food. They embrace a far more cosmopolitan diet than their ancestors, happily snaffling up anything from young iguanas to sea lion afterbirth. Darwin quickly realised that this behaviour

FIGURE 6.3. **The Galápagos hawk.** This vulturous raptor is the top predator in the islands, with a cosmopolitan diet that ranges from young iguanas to sea lion afterbirth. *Reproduced from John Gould, 'Part 3. Birds,' in* The Zoology of the Voyage of H.M.S. Beagle, *ed. Charles Darwin (London: Smith, Elder and Co., 1838–1843), 23.*

might simply be explained by the principle of an animal coming to an island where it could live but finding 'causes to induce great change'. In zoological jargon, this is known as convergent evolution, in which one creature resembles another not through common descent but because the common trait is an effective way to live.

Still, life in the Galápagos is clearly a struggle for this wayward hawk. This much is obvious from the number of birds that fail to breed each year, with many young or inexperienced birds hanging out in non-breeding groups. The death rate in these groups is high, much higher than amongst breeders. This is probably why more than one male and sometimes as many as eight are prepared to attach themselves to a single breeding female. This so-called polyandry is extremely rare in birds, documented in just a handful of 10,000 or so described species. The case of one female and eight males, observed on Santiago in the early 1990s, makes the Galápagos hawk one of the most extreme cases of polyandry in the avian world. Although a female hawk will lay just a couple of eggs a year, which her consorting males can have no guarantee that they've sired, they will still go to the trouble of feeding her and her chicks until they fledge. It's probably worth doing so, just for the increased chance of another year's survival and the possibility of some paternity.

Interestingly, the eradication of invasive mammals across the archipelago over the last few decades (on which more later) has had profound effects upon hawk society. This is most obvious on Santiago, where the Galápagos National Park Service shot more than 17,000 pigs and 70,000 goats between 1998 and 2006. Without a steady supply of carcasses to keep them going, the pool of non-breeding floaters evaporated. Adults too found it harder to survive, and of all breeding groups, the larger ones fared best. In years to come, this is likely to have an effect upon the Galápagos hawk's unorthodox breeding system. With a smaller population, there may be less competition for nest sites and fewer of these male-heavy breeding groups.

If the eradication of goats has rendered the Galápagos hawk harder to see, it has led to a dramatic recovery of other birds like the secretive Galápagos rail, a small ground-dwelling species that wades its way through leaf litter in search of invertebrates like snails, beetles and ants. Back in the 1980s, for instance, a survey on Santiago detected only around twenty individuals, and it looked like the Galápagos rail might suffer the same fate as so many other island-dwelling rails. By 2005, with mammal eradication nearing completion, a similar survey located almost three hundred individuals.

Another surprising sight in the Galápagos is flamingos, a species we more commonly associate with the productive lakes of eastern Africa than an arid landscape like that of the Galápagos. But it just so happens that several brackish lagoons isolated from the ocean are home to a breeding population of the American flamingo. Their movement stirs up the mud, and they use their upturned bills to filter out bacteria, worms and crustaceans. Of all the six different species of flamingo in the world, this one is striking for the psychedelic orange of its feathers. The intensity of this colouration is down to the particular suite of carotenoid pigments contained in the microorganisms that live in the sludge they sift.

There are other wonderful avian inhabitants of the Galápagos. There are the tiny yellow warblers that flit their way through the coastal zones of most islands, nipping up hopping insects that spring from the sand. A recent study shows that as with other birds, there are clear genetic differences from one island to the next, though it looks like the first yellow warblers only reached the archipelago some 300,000 years ago and the different lineages are still in the very earliest stages of speciation. Visitors might catch sight of the Galápagos flycatcher or perhaps one of two species of owl. If they are really lucky, they may glimpse a vermilion flycatcher, the male clothed in a showy suit of red and black. Even more beautiful (in my opinion) is the Galápagos dove. It is pigeon-like in stature, yet stands out with its scarlet legs, terracotta chest and neck, gentle black and white flecks on its wings and icy-blue ring around its eyes.

The Tameness of the Birds

It's possible to get incredibly close to the wildlife in the Galápagos. Anyone who's been to the islands will know just how moving this is. It certainly was for the captain of HMS *Blonde*, George Byron. 'The place is like a new creation,' he wrote of his experience of the islands in 1825. 'The birds and beasts do not get out of our way; the pelicans and sea-lions look in our faces as if we had no right to intrude on their solitude; the small birds are so tame that they hop upon our feet; and all this amidst volcanoes which are burning around us on either hand. Altogether it is as wild and desolate a scene as imagination can picture.'

Darwin had a similar reaction, making a special note of 'the extreme tameness of the birds'. He wrote in his diary, 'The birds are Strangers to Man & think him as innocent as their countrymen the huge Tortoises.' On one day, he prodded a hawk with the muzzle of his gun; the bird fell off its branch. On another, a mockingbird alighted on the edge of a cup (tragically fashioned from the shell of a baby tortoise). Darwin reached out slowly, picked up the shell-cum-cup and lifted it—mockingbird and all—from the ground. On Floreana, he saw a boy by a well with a whip-like stick in his hand. As doves and finches came to drink, the youngster would strike them down, piling their warm bodies into a heap for his dinner. Darwin scratched his head: 'It would appear that the birds of this archipelago, not having as yet learnt that man is a more dangerous animal than the tortoise . . . , disregard him, in the same manner as in England shy birds, such as magpies, disregard the cows and horses grazing in our fields.'

William Beebe found the Galápagos Sally Lightfoot crabs to be an exception: 'They would always sidle out of reach, slowly if I approached gently, or like a scarlet flash if I grabbed quickly,' he wrote in *Galápagos: World's End*. Yet in one spot Beebe found three crabs, 'which in point of fearlessness might have been the Three Musketeers'. As he waded ashore into a secluded cove on Santa Cruz, one particularly large specimen came to meet him. Beebe stood still until the crab—'one of the biggest, his carapace fairly aflame in the sunlight'—had come within reach. Then the naturalist leaned down to rub its shell. The crab 'sank down upon the sand, lowered his eyes into their sockets, and wiggled his maxillipeds ecstatically'. Beebe took all manner of liberties with his new friend, 'lifting one leg after another, raising him from the ground, replacing him, standing him upon his head, and tapping gently upon his hard back'. He concluded, perhaps unfairly, that 'this must be a very ill crab, or an idiot crustacean, or somehow abnormal.' But much to his amazement, when he turned to leave, the tame crab followed.

The importance of this kind of experience to the modern identity of the Galápagos has not been stated forcefully enough. Getting to see a lot of interesting wildlife is one thing. Getting so incredibly close to it and being completely ignored by it is another thing altogether. It gives the concept of 'being at one with nature' a whole new and extremely powerful meaning, one that will sear the Galápagos experience into the mind forever. This certainly seems to have been Beebe's reaction: 'Once we were taught that the earth was the centre of the universe; then that man was the raison d'être of earthly evolution. Now I was thankful to realize that I was here at all, and that I had the great honour of being one with all about me, and in however small a way to have at least an understanding part.'

But it's not, of course, just the birds that are accepting of humans. It's everything else too, and that means the archipelago's iconic reptiles.

Chapter 7. Reptiles

One of the most striking things about the Galápagos is the near complete absence of land-based mammals. Here, it's the reptiles that rule, a group of animals that has rarely achieved such dominance since the days of the dinosaurs.

Darwin was well aware of this curious reptilian ascendancy. 'We must admit that there is no other quarter of the world where this Order replaces the herbivorous mammalia in so extraordinary a manner,' he wrote in the *Journal of Researches*. It was an observation echoed some fifteen years later by Melville in the first of his sketches of 'the Encantadas'. 'Little but reptile life is here found: tortoises, lizards, . . . snakes, and that strangest anomaly of outlandish nature, the *iguana*. No voice, no low, no howl is heard; the chief sound of life here is a hiss.'

The Galápagos is most famous for, and is named after, its gargantuan tortoises, slow and plodding giants that we will encounter in due course. But before we do, we must make our way past Melville's anomalies of outlandish nature, the basking mass of thorny heads, tangled limbs and scaly tails that litters every shoreline in the Galápagos.

Imps of Darkness

Darwin's instinctive response to the marine iguanas was one of revulsion. 'The black Lava rocks on the beach are frequented by large (2–3 ft.) most disgusting, clumsy Lizards,' he wrote of his first encounter on San Cristóbal. 'They are as black as the porous rocks over which they crawl,' he went on, describing them as 'imps of darkness.'

It's more than likely he lifted this delightful phrase from the captain of HMS *Blonde*, George Byron. On Fernandina in 1825, Byron had landed 'among an innumerable host of sea-guanas, the ugliest living creatures we ever beheld'. These reptiles, he wrote, were 'like the alligator, but with a more hideous head, and of a dirty sooty black colour, and sat on the black lava rocks like so many imps of darkness.'

That same year, whilst Darwin was still in shorts, the Zoological Society of London had received a specimen from a contact in Mexico, one that caused herpetologist Thomas Bell much excitement. In a rather poetic contribution to the society's journal, he acknowledged how much progress had been made in establishing the principles of classification, 'probably approaching to the grand plan upon which the animal world was created.' But the bizarre iguana on the table in front of him gave pause for reflection: 'Our knowledge of the natural arrangement must be confessed to be as yet confined to a feeble glimmering of light, the first bright light, as it were, of dawn.'

To Bell, the marine iguana must have been an extraordinary sight to behold and not just because of the incredibly bad taxidermy. He noted the creature's powerful limbs, its flattened tail, its short toes of almost equal length, each sporting a 'remarkably strong, and much hooked' claw. Of all its characteristics, he was most taken by its 'short, obtusely truncated' head and its numerous teeth that were completely unlike those of all other iguanas he'd encountered. From this, he came up with the genus name *Amblyrhynchus*, literally, 'blunt nose'. 'These circumstances,' he concluded, 'evidently indicate some striking peculiarity in its food and general habits,' though he held back on speculating what these might be.

FIGURE 7.1. *Marine iguanas on Española.* The marine iguana is one of the most extraordinary species in the Galápagos, a reptile that has evolved to swim in the sea, diving down to considerable depths to feed on algae. *Reproduced from Walter Rothschild and Ernst Hartert*, Novitates Zoologicae 6 (1899): 7–205.

A decade later, Darwin was the first to get to grips with the striking peculiarity that Bell had predicted. These reptiles swim in the sea, a talent that no other living iguanid has mastered to the same degree. Darwin figured they were probably doing so to feed and set out to find out on what. The simplest thing to do so was to cut one open and take a look in its stomach. To his surprise, he found it 'largely distended with minced seaweed', a strange result but one that he successfully replicated with several other dissections. He could find no sign of any of the bright green and dull red seaweed anywhere on land, so he came to the conclusion that 'it grows at the bottom of the sea, at some little distance from the coast'.

Bell's description had been full of insight. The marine iguana's webbed toes and tail allow it to swim. Its long claws help it to get a grip against the currents and to clamber up the rocks and out of the sea. Its blunt snout and lobed teeth set close to the edge of its jaw are perfect for grazing on the short swards of algae that decorate the rocks.

In the process of catching marine iguanas for dissection, Darwin discovered something else about this peculiar beast: 'that when frightened it will not enter the water'. One iguana in particular bore the brunt of Darwin's inquisition. 'I threw one several times as far as I could, into a deep pool left by the retiring tide; but it invariably returned in a direct line to the spot where I stood.' In spite of the fact that it was obviously comfortable in water, swimming near the bottom 'with a very graceful and rapid movement', no amount of harassment on Darwin's part could induce it to enter the water of its own volition; nor did it attempt to bite him. 'Perhaps this singular piece of apparent stupidity may be accounted for by the circumstance, that this reptile has no enemy whatever on shore, whereas at sea it must often fall prey to the numerous sharks.'

Sharks, or course, do pose some threat to marine iguanas. But more significant is body temperature. Cold-blooded creatures like the Galápagos marine iguana must pay particular attention to this, so it should come as no surprise to discover that their enthusiasm for getting wet changes over the course of each day and from one season to the next. Most marine iguanas do their swimming only when the sun is high and has had a chance to warm them and the sea. In the hot season—from December to May—a marine iguana can afford to spend around an hour a day at sea. In the cool season though—from June to November—the same iguana will only be able to get warm enough to swim for around twenty minutes. We know Darwin was flinging his iguana into the sea in October when the sea temperature is at the low end of its natural range. We do not know what time of day he carried out his experiment, but if it was in the early morning or evening, the iguana's principal concern—far greater than the threat posed by this reptile-slinging Victorian or indeed a shark—would have been to get the heck out of the water.

Darwin also noted that the marine iguanas of Isabela seemed to be larger than those of the other islands. He was not wrong. Isabela is home to the largest recorded specimen, a whopper of an iguana that weighed in at 12 kg, roughly the same as a typical toddler. On Genovesa, by contrast, the largest recorded iguana weighed less than 1 kg. This is partly explained by the upwelling of nutrient-rich waters around Isabela, resulting in longer, more nutritious strands of algae. The different suite of algae also results in differences in colouration from one island to the next.

On any given island, being the biggest can bring big advantages. Large males are able to have more sex than small males. Intriguingly, though, small males are not entirely idle. It takes a male marine iguana around three minutes from penetration to ejaculation. The largest males are tough enough to repel the takeover attempts of other males and stay on top for this length of time, achieving ejaculation in over 95 percent of copulations. Small males, by contrast, are not so successful, getting dislodged by a larger male roughly once in every three mating attempts. Pint-sized males, however, have come up with an intriguing counter-strategy. They prepare what researchers have politely termed an 'ejaculation-in-advance'. It's probably easier just to call it masturbation, though there is a crucial twist: a small iguana holds this ejaculation-in-advance in specialised pouches in his penis, shaving entire minutes off the act of coupling and increasing his chance of transferring sperm to the female before he's booted off.

So big iguanas don't have it all their way. This is especially the case when El Niño comes to town. Then, as you'll recall, those cold, nutrient-rich upwellings falter, and the sea starts to warm, increasing from a surface temperature of around 18°C in a normal year to a maximum of 32°C during an El Niño. This might make swimming less of an issue for the cold-blooded iguanas, but there's not much point in taking a dip if there's no food to be had. In these conditions, the reptiles' favoured red and green algae dies off, and their only option is to turn to the far less digestible brown algae that thrives in its place.

Thirty years ago, during the extreme El Niño of 1982–1983, researchers first witnessed the devastating impact this had on marine iguanas. Around two out of every three animals starved to death. The very youngest perished, probably because they had not yet honed their foraging skills, and so did the oldest and largest.

In the following decade, researchers noticed something even more startling: their measurements of body length seemed to suggest that individual iguanas were getting smaller from one season to the next. At first they thought there had to be some mistake, but in time it became clear that this is exactly what happens. Adult iguanas have a flexible skeleton, shrinking it in years when food is scarce—sometimes by more than 5cm—and growing it again when supplies recover. Why would they do this? It certainly looks like a nifty adaptation to survive. During the 1992–1993 El Niño, the researchers found that the animals able to shrink the most had a better chance of survival.

The marine iguana has other extraordinary adaptations too. Darwin noted a member of the *Beagle* crew attempting to drown one, weighting it down, attaching a line and tossing it overboard. There is no explanation why the seaman didn't just bop the unfortunate reptile on the head, but Darwin was intrigued to find that 'when, an hour afterwards, he drew up the line, it was quite active'.

In the early 1960s, a pair of zoologists carried out something similar, only in a laboratory at the University of California, Los Angeles. The idea of shipping marine iguanas to the United States and then forcibly submerging them for half an hour in waters of varying temperature might seem fairly brutal (it would certainly stand no chance of getting past a twenty-first-century ethics committee), but it did reveal how the heart rate of these reptiles plunges when they dive, dropping from around forty to about ten beats per minute.

Marine iguanas are clearly also able to tolerate considerable pressures, with some animals recorded descending to depths of 30m. As they inevitably swallow a lot of seawater along with their algal nibblings, they also need a way to excrete all that salt. Reptiles do this by way of a specialised salt gland located above and between their eyes. When a marine iguana snorts, firing out droplets of what appear to be snot, it is in fact ejecting excess salt. This messy practice helps explain why the face of a marine iguana often appears to be spattered with a whitish bloom.

Land Iguanas

Whilst marine iguanas are an extremely common sight in the Galápagos, the same cannot be said for their terrestrial cousins, the less numerous and more reclusive land iguanas. These come in three principal flavours: the most widespread of these, *Conolophus subcristatus*, can be encountered on several islands, including Santa Cruz, Baltra, South Plaza,

PLATE I. In 1825, American explorer Benjamin Morrell witnessed the eruption of Fernandina at terrifyingly close quarters. 'The heavens appeared to be in one blaze of fire, intermingled with millions of falling stars and meteors,' he wrote. Fernandina, still the most active volcano in the Galápagos, is captured here mid-eruption in April 2009. Copyright © Paula A. LeVay.

PLATE 2. The Galápagos Marine Reserve, formally established in 1998, is one of the world's largest protected marine areas, filled with life from microorganisms to behemoths like this humpback whale. Copyright © Jo Anne Rosen.

PLATE 3. With the power of flight, seabirds were quick to colonise the Galápagos, using the bare rocks as nesting bases and the ocean as a larder. The blue-footed booby is one of the most charismatic seabirds in the islands. Copyright © Catherine Dobbins d'Alessio.

PLATE 4. Brown pelicans drift through the Galápagos landscape. In the background, it's possible to discern the different ecological zones that characterise these islands: the coastal zone is a magical habitat boasting mangroves and other salt-tolerant plants, and beyond is the arid, scrubby hinterland that Charles Darwin felt contained 'wretched-looking little weeds'. At greater altitudes, pictured here in the distance, there is enough rainfall to sustain a more effulgent flora. Copyright © Andy Teucher.

PLATE 5. There is only one bee in the islands: the Galápagos carpenter bee. It seems happy to feed from an impressive array of flowering plants, which may be why it got a hold in the islands whilst other species, if they tried, failed. In the Galápagos, it does not pay to be too fussy. Copyright © Carol S. Hemminger.

PLATE 6. With plants bedding down and insects finding their niche, land birds were in a position to call the Galápagos home. Most of them, like Darwin's finches, are rather ordinary to look at. But not so the male vermilion flycatcher. Copyright © Katsunori Namikata.

PLATE 7. In the Galápagos, it's the reptiles—a group of animals that has rarely achieved such dominance since the days of the dinosaurs—that rule. The marine iguana, which Herman Melville described as 'that strangest anomaly of outlandish nature', lives in the coastal zone, a basking mass of thorny heads, tangled limbs and scaly tails. Copyright © Geoff Goodyear.

PLATE 8. The giant tortoise, the dominant herbivore that gave the islands their name, is one of the most extraordinary sights in the Galápagos. There is increasing evidence that these creatures act as 'ecosystem engineers', exposing soil for plants to take root, opening up dense vegetation, dispersing the seeds of the plants they eat and possibly even helping them to germinate. Copyright © Kathy Reeves.

Isabela and Fernandina; then there's a pale variety, *Conolophus pallidus*, found only on the tiny island of Santa Fé; and a third species is now recognized, the mysterious pink land iguana *Conolophus marthae*, found only on Isabela's Wolf Volcano.

The first land iguana to come to the attention of western zoologists—a specimen of *Conolophus subcristatus*—happened to turn up in the museum of Boulogne-sur-Mer in France. Once Paris-based herpetologist Gabriel Bibron got wind of this, he became the first to describe it in 1837, just before Thomas Bell (still at the Zoological Society of London) got his hands on Darwin's *Beagle* specimen. Bibron and his colleague André Duméril had little doubt of its similarity to the marine iguana and paired them off as members of the same genus, *Amblyrhynchus*.

Darwin independently arrived at the same conclusion about the relationship between this new land iguana and the more abundant seagoing variety. 'This animal clearly belongs to the same genus as the last—it being a terrestrial, whilst the other is an aquatic species,' he wrote in his *Zoology Notes* during the voyage. 'It would appear as if it had been created in the centre of the archipelago, and thence has been dispersed only to a certain distance.'

Today, the marine and land iguanas are assigned to different genera, *Amblyrhynchus* and *Conolophus*, respectively. This acknowledges their divergent habits but ignores what Bibron and Duméril and Darwin and Bell would now recognise as their shared evolutionary history.

In 1983, a couple of immunologists took a look at the similarities and differences between these two iguanid forms. Based on an analysis of blood proteins, they estimated the two lineages had been going their separate evolutionary ways for somewhere between 15 and 20 million years. Over this amount of time, they reasoned, there would have been sufficient genetic change to account for the different ways in which the marine and land iguana proteins behaved in the lab. Subsequent studies, each using the genetic techniques du jour, have all produced a similar result, with the origin of these species usually placed at around 10 million years ago or more.

At first glance, this poses a bit of a puzzle. If the oldest island in the Galápagos erupted less than 4 million years ago, where is this speciation supposed to have taken place? It could, of course, have occurred on mainland South America, but there are a couple of problems with this scenario. First, iguanas would have to have reached the Galápagos twice, once for the ancestors of marine iguanas and once for the ancestors of land iguanas. Second, there's not a shred of evidence—either fossil or living—that anything like the Galápagos iguanas has ever lived on the continent.

Thankfully, the geological underpinnings of the Galápagos, which we explored in Chapter 1, collapse the conundrum. Iguanas only found their way to the Galápagos once, giving rise around 10 million years ago to two different forms that eventually became the marine and land iguanas we see today. But this colonisation and subsequent divergence took place on islands we no longer see, ones that have moved off the Galápagos hotspot, cooled and contracted back beneath the waves to assume their position on the submerged Carnegie Ridge.

'A Singularly Stupid Appearance'

As with the marine iguanas, Darwin was not won over by the looks of the land variety: 'From their low facial angle they have a singularly stupid appearance,' he wrote. 'In their movements they are lazy and half torpid.' On Santiago, where he found them ranging from the lower arid zone up into the central damper regions of the island, he decided to cut open a few and found them 'full of vegetable fibres and leaves of different trees, especially of an acacia.' On the lower slopes, they'd been particularly busy burrowing, so much so that he had trouble finding a patch of land on which to pitch his tent. 'When walking over these lizard-warrens, the soil is constantly giving way, much to the annoyance of the tired walker.' Presumably, the iguanas were also peeved by this unwaveringly investigative human: 'I watched one for a long time, till half its body was buried; I then walked up and pulled it by the tail; at this it was greatly astonished, and soon shuffled up to see what was the matter; and then stared me in the face, as much as to say, "What made you pull my tail?"' Darwin did not proffer an answer to this perfectly reasonable question; he was too busy subjecting the iguanas to more misery: if you hold one and plague it with a stick, he discovered, 'it will bite very severely'; if you put two together on the ground, 'they will fight, and bite each other till blood is drawn'; if you toss a bit of succulent cactus to them, they will 'seize and carry it away in their mouths, like so many hungry dogs with a bone'.

Although Darwin's actions can only have had a very minor impact on the land iguanas of Santiago, the impact of other mammalian feet (mainly those of donkeys, pigs and goats) on their burrows and wider habitat probably accounts for the sad fact that none is left on Santiago today, and their numbers are much reduced elsewhere.

Of the ones Darwin saw (*C. subcristatus*), he described their belly, legs and head as being 'Saffron' and 'Dutch orange' in colour. The land iguanas of Santa Fé (*C. pallidus*) are judged to be a different species owing to their lighter skin tone and a few other characteristics. More recently, in

FIGURE 7.2. **Land iguana.** *Conolophus subcristatus* is the most common of three species of land iguana in the Galápagos. *Reproduced from Thomas Bell, 'Part 5. Reptiles,' in* The Zoology of the Voyage of H.M.S. Beagle, *ed. Charles Darwin (London: Smith, Elder and Co., 1838–1843), 22.*

1986, rangers from the Galápagos National Park Service discovered that Wolf Volcano on Isabela is home to a pink variety, but with the pressures of other work, it took almost twenty years before a dedicated expedition went in search of this by then nearly mythological beast. In 2005, park wardens found four pink iguanas. The following year, they located a further thirty-two. These expeditions resulted in the formal description of a new species (C. *marthae*).

Apart from its striking colouration, this new species has several other notable differences from its yellow and pallid cousins. Whereas C. *subcristatus* and C. *pallidus* both have pretty prominent spines on their heads and down their backs, these are almost absent on C. *marthae*. In addition, each form has its own distinctive head-bobbing routine, a display that the land iguanas of Santiago appear to have performed for Darwin in 1835. 'When attentively watching an intruder they curl their tails, & raising themselves as if in defiance on their front legs, vertically shake their heads with a quick motion,' he wrote in his *Zoology Notes*. This head-bobbing display is a common feature in iguanas, used in territoriality and courtship.

But the result of the genetic analysis is the most intriguing, strongly suggesting that C. *marthae* has descended from an ancient lineage that split away from the other land iguanas some 5.7 million years ago. If this is the case, it's weird that C. *marthae* is confined to Wolf Volcano and is not found on any of the older islands.

One possibility is that the arrival of humans and other introduced species has, for some reason, taken a disproportionate toll on the pink iguana, though given that the pink iguana does not show up in any of the many historical accounts of Galápagos fauna, this seems a little unlikely. Another possibility is that it's been usurped by C. *subcristatus*. Whatever the explanation for its incredibly confined distribution (it's thought to occupy no more than 25 km²), there's a serious danger that it could soon disappear altogether. Only one in every three of the pink iguanas discovered during the 2006 expedition was female, and there were no signs of juveniles. A more recent expedition, in 2012, found evidence that reproduction is still occurring, with a youngster estimated to be around four years old, but clearly the situation is touch-and-go. If the Galápagos National Park Service gets an inkling that the population is about to vanish, it may be time to think about attempting to coax some of the remaining individuals to breed in captivity.

Giant Tortoises

The taxonomy of the Galápagos giant tortoises has always been something of a mess. At first, it was assumed that they were simply the same as those found on the Mascarene Islands in the Indian Ocean, perhaps transported to the Galápagos by thoughtful sailors. Then it was agreed that they were a different species. By 1835, there were those who had guessed correctly that each island harbours its own peculiar variety. We know this because Nicholas Lawson (the vice governor of the islands) said as much. If presented with a random tortoise, he bragged to Darwin, he could confidently assign it to its island of origin simply by looking at the shape of its shell.

Darwin clearly didn't pay much attention to Lawson's claim. If he had, he would surely have made more effort to collect giant tortoises and spent less time feasting on tortoise meat and tossing their empty shells into the Pacific. But at some point en route to Tahiti, he seems to have sat up, scribbling a note into his diary to remind him of the conversation with the vice governor. In the first edition of his famous *Journal of Researches*, published in 1839, Darwin mentioned this encounter in passing only. By the time the second edition appeared in 1845, he'd realised its profound importance. The following excerpt is one of the most frequently cited passages from any of Darwin's many books and as close to a eureka moment as you're likely to get:

> I have not as yet noticed by far the most remarkable feature in the natural history of this archipelago; it is, that the different islands to a considerable extent are inhabited by a different set of beings. My attention was first called to this fact by the Vice-Governor, Mr. Lawson, declaring that the tortoises differed from the different islands, and that he could with certainty tell from which island any one was brought. I did not for some time pay sufficient attention to this statement, and I had already partially mingled together the collections from two of the islands. I never dreamed that islands, about 50 or 60 miles apart, and most of them in sight of each other, formed of precisely the same rocks, placed

under a quite similar climate, rising to a nearly equal height, would have
been differently tenanted.

Over the course of the nineteenth century, a succession of naturalists
came to the Galápagos intent on providing a more objective account of
these differently tenanted islands. In the wake of the California Academy of Sciences expedition to the Galápagos in 1905 and 1906, it fell to
herpetologist John Van Denburgh to assimilate all the opinions expressed
in preceding decades. In a two-hundred-page manuscript, he recognised
fifteen different species of Galápagos tortoise, two of which he judged
had already gone extinct. Van Denburgh's classification has largely stood
the test of time. Today we recognise fourteen different species of Galápagos giant tortoise, of which four are now extinct. In general, the rule of
thumb is one island, one tortoise, but on the biggest island, Isabela, each
of the five major volcanoes has its own unique form.

The most obvious trait to differ from one island to the next is the
shape of the shell. It's the tortoise equivalent of the finch's beak. At one
extreme there are species with a conventional 'domed' carapace. At the
other are the 'saddlebacks', tortoises with an exaggerated saddle-like furl
at the front. So striking is this difference that David Porter, captain of
the US frigate *Essex*, made a note of it two hundred years ago. He judged
the tortoises on Santiago to be 'a species entirely distinct' from those on
Española. On Santiago, he found the tortoises 'round, plump, and black
as ebony, some of them handsome to the eye'. On Española, he found
the shell 'elongated, turning up in the manner of a Spanish saddle'.

Since most tortoises in the world have a domed shell, it's reasonable
to assume that this is the default position and the saddle the exception.

FIGURE 7.3. **Right, *Galápagos giant tortoise*.** It's thought there were once
at least fourteen different species of giant tortoise in the Galápagos, of which
only ten now remain. These come in two principal forms—domed and saddle-
backed—both of which are depicted in this nineteenth-century engraving. *Reproduced from Charles Frederick Holder,* Charles Darwin: His Life and Work *(New
York, London: G. P. Putnam's Sons, 1892).*

ELEPHANT TORTOISE, GALAPAGOS ISLANDS.

So what's the point of such an extraordinary shape? The conventional wisdom is that it helps survival, particularly on flattish islands where there is a distinct dearth of low-lying vegetation. Santiago's main peak reaches just over 900m above sea level and is clothed in dense, lush vegetation, and a tortoise can get along nicely with a fairly conventional shell shape. On an even higher volcano like Isabela's Sierra Negra, the tortoises are more domed still.

Española, however, sits in the eroded south-east of the archipelago and reaches only 200m at its highest point; in the absence of highlands and plentiful food, a domed tortoise just isn't going to survive. A saddleback tortoise, by contrast, has a far higher reach. It has not only the obvious opening in the front of its shell but also a longer neck and legs to boot. A large saddleback on tiptoe with its neck at full stretch is able to lever its head almost 2m above the ground—an impressive sight indeed. As tricky as it might be to hold this extensive pose, it gives the tortoise access to lofty sources of nutrition like the much coveted fruits of the *Opuntia* cacti.

On San Cristóbal, the giant tortoises have a rather open-ended shell too, and when Darwin encountered his first specimens on this island, one was munching on a cactus. 'Surrounded by the black Lava, the leafless shrubs & large Cacti, they appeared most old-fashioned antediluvian animals; or rather inhabitants of some other planet,' he wrote in his diary. Once he'd had a chance to observe other tortoises on other islands, he recognised that those confined to the arid zone eat different things to those in the highlands. 'The tortoises which live on those islands where there is no water, or in the lower and arid parts of others, feed chiefly on the succulent cactus,' he wrote in his *Journal of Researches*. 'Those which frequent the higher and damp regions, eat the leaves of various trees, a kind of berry (called guayavita) which is acid and austere, and likewise a pale green filamentous lichen . . . that hangs in tresses from the boughs of trees.'

On Santiago, he saw how they carved their way through the undergrowth, leaving 'well beaten roads' that appeared to lead to springs in the highlands. He climbed on top of one tortoise and found it 'so strong as easily to carry me' and tracked another—slowly—along one of these ascending tracks, estimating its speed at one-sixth of a mile per hour.

'When they arrive at the Spring, they bury their heads above the eyes in the muddy water & greedily suck in great mouthfulls, quite regardless of lookers on.'

Recent research has added to Darwin's early observations. It makes intuitive sense that a giant herbivore should have an impact on its surroundings, exposing soil for plants to take root, opening up dense vegetation, dispersing the seeds of the plants it eats and possibly even helping them to germinate. But what's the evidence that they really do act as so-called ecosystem engineers?

The tortoise corrals at the Charles Darwin Research Station are a good place to start. They ring-fence not only the tortoises but also the plant communities growing there. So a comparison of enclosures—some with tortoises and some without—should give an idea of what impact these animals have on the vegetation. By munching and trampling, tortoises reduce the density of small cactuses, especially in the vicinity of the biggest cacti. Although this might sound like a bad thing for those cacti that get eaten, the reintroduction of giant tortoises to Española (about which we will learn more in the Chapter 9) shows that the reptiles seem to be spreading seeds further afield with a net benefit to the cactus population as a whole.

The obvious way to get a better feel for this is to look at poo. Having teased apart more than one hundred piles of tortoise dung, researchers calculated that a deposit commonly contains hundreds of seeds, often from more than just one species. Although tortoises don't move particularly fast, nor does the passage of food through their guts. In fact, it can take almost a month for a bit of food to pass from one end to the other, though twelve days is more typical. In this time, even the most sluggish of tortoises will have moved some distance, typically more than 500m and in one instance more than 3 km, from the parent plant.

Perhaps the biggest test of the giant tortoise's role as an ecosystem engineer is taking place on Pinta. This island has not had a tortoise since 1972 (when the last-known member of the species, Lonesome George, was taken off the island and into captivity). In late 2009, the Galápagos National Park Service took the bold decision to release thirty-nine *mezclados* onto the island, tortoises of mixed ancestry that had been living out their lives in the confines of the Charles Darwin Research Station.

The purpose of this introduction was not to get a breeding population back on Pinta—the chosen *mezclados* were all sterilized before release—but to find out how adult tortoises adapt to being let loose on a strange island and to study their impact on the vegetation. All of them were kitted out with either radio or satellite transmitters, allowing their movements to be documented simultaneously and from afar. Such studies help to build up a picture of what makes a tortoise tick, how it impacts its world and where its vulnerabilities lie.

Another rich seam of insights into the Galápagos tortoises has come from the study of their DNA. Since the late 1990s, geneticists at Yale University have been intent on characterising the genetic differences from one species to the next. As with the iguanas, one of the first uses of these data was to get an idea of where the Galápagos tortoises came from and when. The discovery that their closest living relative is the comparatively small Chaco tortoise that inhabits arid regions of central South America suggests that the ancestors of giant tortoises had to have come from the mainland. It also suggests that these two lineages—one leading to the Chaco and one to the Galápagos—went their separate ways somewhere between 10 and 20 million years ago and the first Galápagos tortoises reached either San Cristóbal or Española around 3 million years ago.

It's reasonable to ask what the Galápagos ancestors were doing between 10 and 3 million years ago? One possibility is that they were hanging out on the continent; there are giant tortoise fossils there that date to around 2 million years ago. The alternative is that, rather like the ancestors of Galápagos iguanids, they have been living on islands all this time, repeatedly hopping from sinking islands to occupy the Galápagos hotspot's latest volcanic issue.

My instinct is that tortoises have been island hopping. As with the iguanas and presumably much else besides, the tortoises are likely to have moved through the archipelago on the prevailing currents from the south-east to the north-west. This requires the occasional tortoise to find itself at sea; given the length of time we're dealing with—hundreds of thousands of years—this doesn't seem implausible. The tortoise must also avoid sinking; on the few occasions a giant tortoise has been observed at sea, it's bobbed along tolerably well for a time, head above water and limbs cycling in an awkward attempt at propulsion. There must also be an island to receive it, and it is not long after the eruption of each new island that tortoises appear to have settled.

The story is probably remarkably similar for other reptiles, like geckos and snakes, though nobody has looked. One group researchers have investigated is the lava lizards of the Galápagos. Scientists formerly recognise seven species, but there are probably at least a couple of others. The genetic evidence suggests that two independent colonization events best account for their current distribution. One group of lizards appears to belong to a lineage that separated from continental South American lizards several million years ago. As with marine iguanas and giant tortoises, they might have occupied islands that have since slipped beneath the waves, finally reaching Española and using it as a stepping-stone to other islands to the west as and when they appeared. Another group of lizards took a similar course, arriving on San Cristóbal a few million years ago, from where they seem to have launched to Marchena.

So, as with everything else in the Galápagos, evolutionary change has engendered an impressive diversity of reptilian life, giving the illusion that that these cold-blooded creatures have come to dominate these islands. But this ignores the arrival of *Homo sapiens*. Over the course of the final three chapters, we will see that the Galápagos has changed this species too.

Chapter 8. Humans: Part I

Compared to the prickly pear opuntias, the wrasse, the boobies, the weevils, finches and iguanas, humans came late to the Galápagos. But like all of the settlers before them, humans reached these islands by accident rather than design, swept out from the South American coast—against their will—on what we now call the South Equatorial Current. They also did so by rafting. What, after all, is a Spanish galleon if not a mat of (admittedly rather special) vegetation?

It is possible that this first ship was more like a canoe and belonged to the pre-Columbian people of modern-day Ecuador or Peru. There are certainly those in both countries who claim their own Amerindians were the first to reach the Galápagos. If they were, however, there is—as yet—no compelling evidence, and the bishop of Panama's fleeting visit in 1535 is the first to be documented.

On the Map

Once safely back in Lima, the bishop wrote a report of his Galápagos misadventure to King Charles I of Spain. 'The Lord fill Your Sacred Majesty with holy love and grace for many years and with the conservation of your realms and an increase of other new ones, as I hope,' he wrote, though there was no recommendation that the Galápagos should be one of them. 'I am your most true servant and subject and perpetual Chaplain, who kisses your royal feet and hands.'

It is from the accounts of the bishop and those who followed that the islands got their name. In the Geography and Map Division of the Library of Congress in Washington, DC, there is a stunning sixteenth-century chart of the South American coastline from Guatemala to Peru with significant landmarks scratched onto a rectangle of cow hide in black, blue, green and red inks. It's easy to make out modern-day Panama City, Bogota, Quito and Lima, urban centres all marked with little cartoon-like palaces. All the way along the coast, offshore islands—obvious hazards to the nautical folk who would have been using this chart—are flagged up in red. Further out to sea, all alone, is a collection of twelve elongated blobs, and projecting downwards in elegant script are the unmistakable words *ys de galapagos*.

It's often said that *galápago* was the medieval Spanish word for 'saddle' and that the islands were therefore named after the exaggerated saddle-shaped shells of several of the tortoise species. In fact, as historian John Woram has made clear, it was the other way round. *Galápago* meant 'tortoise', and the *silla-galápago* (which turned up as an equine accessory in Spanish literature in the nineteenth century) was a tortoise-shaped saddle.

So the *ys de galapagos* are the 'islands of the tortoises'. It's an appropriate name. The accounts of all early visitors reveal that the giant tortoises made a deep impression. Sadly, these creatures served another, more utilitarian purpose. The palpable excitement of those who dined out on their flesh persuades me that these creatures must taste good. For the next three centuries, the principal reason for setting foot on the Galápagos was to take stock of these palatable reptiles.

Tasty Tortoises

One of the first to wax lyrical about the taste of Galápagos tortoises was the buccaneer William Dampier. 'They are extraordinary large and fat; and so sweet, that no Pullet eats more pleasantly,' he wrote in 1697. James Colnett (he of the British whaler HMS *Rattler*) sung the praises of their 'excellent broth'. Amasa Delano, a US Navy captain who described his visit in 1801, felt that 'their flesh, without exception, is of as sweet and pleasant a flavour as any that I ever eat'. In 1813, another US Navy captain, David Porter, was even more gushing: 'Hideous and disgusting as is their appearance, no animal can possibly afford a more wholesome, luscious, and delicate food than they do,' he wrote. Although Darwin judged tortoise meat to be 'indifferent food', he conceded that when roasted on the bone in the manner that South Americans roast their beef, 'It is then very good.' Moreover, 'Young Tortoises,' he claimed, 'make capital soup.'

It was not just the meat that excited these men. According to Dampier, one party that had feasted on tortoise meat for some three months 'saved sixty Jars of Oyl', which once back at sea 'served instead of butter to eat with doughboys or dumplings'. Delano gave a more accurate sense of how much fat could be collected from one animal. Besides that used to fry up the meat, 'it was common to take out of one of them ten or twelve pounds of fat,' he wrote. 'This was as yellow as our best butter, and of a sweeter flavour than hog's lard.' Porter concurred. The fat, he found, 'does not possess that cloying quality, common to that of most other animals' and 'furnishes an oil superior in taste to that of the olive.'

The giant tortoises proved to be a handy source of liquid refreshment too. 'They carry with them a constant supply of water, in a bag at the root of the neck, which contains about two gallons; and on tasting that found in those we killed on board, it proved perfectly fresh and sweet,' wrote Porter.

There is, however, another, less sapid reason why the Galápagos tortoises suffered more than most: their cold-blooded nature means they can drop their metabolism, and with an inbuilt supply of water, they are able to survive up to a year without food or water. In a world before

freezers, giant tortoises were a godsend. 'They were piled up on the quarter-deck for a few days . . . in order that they might have time to dis-
charge the contents of their stomachs,' explained Porter. Then he had
his men carry them below deck and stored 'as you would stow any other
provisions'. Writing about his experience of the Galápagos in 1825, Benjamin Morrell asserted that a hoard of tortoises would provide 'fresh provisions for six or eight months', protecting the men from scurvy into the
bargain.

This combination of tastiness and hardiness resulted in tortoise
slaughter on an absolutely staggering scale. Privateers like Dampier,
whalers like Colnett, naval officers such as Delano and Porter, and
explorers like Morrell weren't just hungry—they were very, very greedy.

The collection of tortoises was something of a military operation. 'All
hands employed in making belts to go after terpen,' reads the entry of
one whaling logbook. 'Four boats were dispatched every morning . . . and
returned at night, bringing with them from twenty to thirty [tortoises]
each,' wrote Porter of a particularly profitable visit to Santiago in 1813.
'In four days we had as many as would weigh about 14 tons on board,
which was as much as we could conveniently stow.' This would have
been just shy of five hundred animals. On another occasion he recorded
'getting on board between four and five hundred' from Floreana.

Whilst such hauls seem to be the exception rather than the rule,
most vessels would typically take dozens and often hundreds. In an anal-
ysis way ahead of its time, director of the New York Aquarium Charles
Haskins Townsend sat down in the 1920s to pore over the logbooks of
American whaling vessels and found several similarly startling harvests.
The crew of one vessel, for instance, spent just five days on Española and
came away with 335 animals. Another took nine days to reap 350 from
Floreana. A third took as many days to secure 240 tortoises from San
Cristóbal. There is the odd document that captures how backbreaking
this must have been for the sailors. 'We got about 250 altogether which
cost us much trouble,' wrote one. Another confessed to being 'tired oute'
by the exertion. Yet another still found himself 'intirely exhoisted'.

By the time he'd finished, Townsend had extracted tortoise tallies
from the logbooks of seventy-nine American whalers that had made
189 collecting trips in the Galápagos between 1831 and 1868. In just

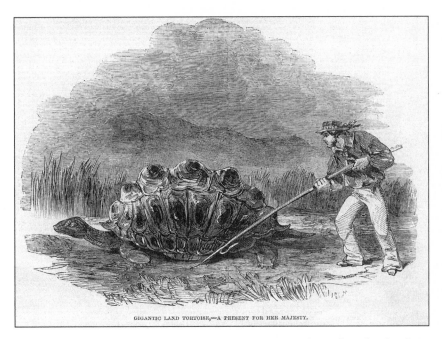

GIGANTIC LAND TORTOISE,—A PRESENT FOR HER MAJESTY.

FIGURE 8.1. *The Galápagos giant tortoises tasted good.* It's thought that buc-caneers and whalers may have eaten their way through several hundred thou-sand giant tortoises. *Reproduced from the* Illustrated London News, *13 July 1850, courtesy of John Woram.*

thirty-seven years, these hunters had removed a minimum of 13,013 giant tortoises from the islands. As this was only a fraction of the Amer-ican fleet and captured none of the exploitation carried out by British whalers, Townsend knew it was a gross underestimation of the toll. 'It would be within safe limits to credit American whalers with taking not less than 100,000 tortoises subsequent to 1830,' he concluded. We can never really know the full extent of the devastation, but factoring in the activities of the British whaling fleet and the likes of Porter, this phase in the history of the Galápagos could well have seen the extraction of 200,000 giants. Or more.

Whilst he was about it, Townsend did the same for fur seals. The few snippets of evidence he could assemble suggested sealers had taken at least 20,000 animals from the archipelago between 1870 and 1882. 'This is of course a trivial number as compared with the total catch made

FIGURE 8.2. *The sperm whale.* Sperm whales were big business in the nineteenth century, and the Galápagos was one of the best whaling grounds around. *Reproduced from Charles Nordhoff,* Whaling and Fishing *(Cincinnati: Moore, Wilstach, Keys & Co., 1856).*

during the period, the records of which are not available,' he wrote. By the time he first visited the Galápagos in 1888, only a few fur seals remained.

By then, the whale population had been decimated too. In the days before the discovery of petroleum in the 1850s, whales had been of enormous commercial value. Oil refined from whale's blubber had provided the most common source of fuel for artificial lighting; it was used to grease up factory machinery and clocks; whalebones were fashioned into canes, corsets and ribs for umbrellas. The sperm whale was of particular value owing to the pearly, ejaculate-like fluid (hence called spermaceti) contained in an unusual cavity in its rectangular head (the spermaceti organ). There's still debate over what purpose this cavity serves the whale: it could be used to control buoyancy; it might be used as a battering ram to stun its prey; or, more likely, it serves some complex acoustic function. The whalers, however, were interested in none of this. Once bailed out from the sperm whale's head and processed, the spermaceti yielded a particularly clarified product suitable for making the finest of candles and the most coveted of cosmetics. In around 1820, when the whale industry was in full swing, spermaceti oil was worth twice as much as bog-standard blubber oil.

In comparison to the decimation of the giant tortoises, fur seals and whales, the decision of humans to live in the Galápagos from the nineteenth century onwards seems innocent enough. But the way in which this settlement played out over the course of the next two centuries is nothing short of remarkable.

Potatoes and Pumpkins

The early efforts to colonise the Galápagos were slow and fraught. The first person to call the Galápagos home is thought to have been an eccentric Irishman called Patrick Watkins, marooned on the island of Floreana in 1806. 'The appearance of this man, from the accounts I have received of him, was the most dreadful that can be imagined; ragged clothes, scarce sufficient to cover his nakedness, and covered with vermin; his red hair and beard matted, his skin much burnt, from constant exposure to the sun, and so wild and savage in his manner and appearance, that he struck every one with horror,' wrote Porter. In the years to follow, Porter's second-hand account spawned several third-hand versions of the Watkins story, most notably Herman Melville's ninth sketch of the islands. 'He struck strangers much as if he were a volcanic creature thrown up by the same convulsion which exploded into sight the isle.'

Watkins lived in a hut around a mile inland from what is now Puerto Velasco Ibarra, the small settlement that opens onto the appropriately named Black Beach on the west coast of Floreana. Porter was fascinated by this character, describing how he managed to survive in solitude, growing his own vegetables and trading them with passing ships in exchange for rum. 'We have seen, from what Patrick effected, that potatoes, pumpkins, &c., may be raised of a superior quality, and with proper industry the state of these islands might be much improved.' Few people had Porter's vision that the Galápagos might be 'improved'. Yet within a couple of decades, a man with a plan did come forward. Louisiana-born José María Villamil had fought hard against Spanish occupation of South America and is one of Ecuador's founding fathers. Upon Ecuador's gaining independence in 1830, Villamil urged the country's first president, General Juan José Flores, to stake a claim on the Galápagos, proposing to set up a small colony to harvest *Roccella tinctoria* (an abundant lichen and good source of the valuable purple dye orchil) for export to the mainland. Flores agreed, and Villamil became the first governor of the island. An advance party of a dozen men reached Floreana in January 1832, and a small ceremony on 12 February formally marked Ecuador's acquisition of the archipelago. This is Galápagos Day, a public holiday in the islands that rather wonderfully happens to coincide with Darwin's birthday.

One of the first reliable accounts of Villamil's endeavours comes from Jeremiah Reynolds, a US naval officer on board the frigate *Potomac*. This reveals, in no uncertain terms, the power that humans have to transform a landscape. Within a year of Villamil's arrival in October 1832 with around one hundred men at his disposal, Reynolds judged that 'the productions of the island are sufficient for several hundred additional inhabitants'. The luxuriant centre of the island boasted soil that 'may be cultivated from January to December, one crop following another in rapid succession; moistened in summer by continued and heavy dews, and by rains in winter'. From a vantage point in the highlands, Reynolds looked down upon Asilo de la Paz ('Refuge of Peace'), 'no less than fifty little *chacras*, or farms, with nearly an equal number of houses'. At a small party held for the visitors, the only beverage on offer was water drawn from the 'Governor's Dripstone' and other freshwater sources that spring from the mountainside. It looked like Villamil's venture was to be a great success, something Reynolds put down to his strict no-alcohol policy.

By the time HMS *Beagle* appeared on the horizon a couple of years later, Villamil's settlement had roughly doubled in size. With his boss absent, Vice-Governor Lawson (he with the interest in tortoise morphology) led Robert FitzRoy and Charles Darwin from Black Beach up to the new settlement in the highlands. Both men were impressed by its productivity. 'Surrounded by tropical vegetation, by bananas, sugar canes, Indian corn, and sweet potatoes, all luxuriantly flourishing, it was hard to believe that any extent of sterile and apparently useless country could be close to land so fertile,' wrote FitzRoy.

Although Villamil's fledgling enterprise folded within a couple of years and there's little left of the original settlement, the wider impact of this agricultural enterprise is only too obvious. In a matter of years, the seeds that would undermine Floreana's ecological integrity had been set.

The impact of introduced mammals is even more obvious. In his summary of the efforts at farming on Floreana, Reynolds also noted that the lower, drier portions of the island would be 'good for raising hogs, goats, &c'. There are now thirty species of vertebrates known to have been introduced to the Galápagos by humans, and most of them have been very bad news. Darwin appreciated the impact that aliens could have on native species, noting 'what havoc the introduction of any new beast of prey must cause in a country, before the instincts of the aborigines become adapted to the stranger's craft or power'. At the last count, there were also 536 introduced invertebrates (at least), several of which—like the little fire ant considered in the next chapter—have had a highly destructive influence.

Invasive plants are a lot harder to demonise than rats and insects. Indeed, the list of plants that humans have introduced to Floreana and the other inhabited islands in the Galápagos (870 species, according to the latest reckoning) gives off a luxuriant scent: guava, hill raspberry, quinine, rose apple, avocado, orange, lemon, lantana, angel's trumpet, bamboo, balsa, and so on. But the damage caused by these species is tremendous owing to the simple fact that the only place suitable for their cultivation is the damp, species-rich highlands (which explains why Villamil set up camp in the centre of Floreana rather than on the coast and why the early settlers on each of the other inhabited islands were quick to turn the most productive humid zone over to the plough).

A recent study of satellite images suggests that over half of the highland area on the four inhabited islands (almost 300 km² in total) has been completely transformed by agriculture. On Floreana, some of this habitat is recovering after the removal of its large invasive mammals, but the situation on Santa Cruz and San Cristóbal is not so good. On San Cristóbal, for instance, more than 95 percent of the highland habitat is seriously degraded as a result of the nearly continuous presence of humans for more than 150 years. This began in the 1850s, when another entrepreneur, Manuel Cobos, set up a commune called 'El Progreso' with the intent of farming lichen (just as Villamil had attempted on Floreana). They ended up growing sugar cane and coffee, but the project ended in disaster, the (largely convict) workforce rising up to kill its punitive foreman in 1904.

FIGURE 8.3. *El Progreso in the highlands of San Cristóbal.* Entrepreneur Manuel Cobos created this settlement in the 1850s, where his largely convict workforce tended plantations of sugar cane and coffee. *Reproduced from the NOAA's Historic Fisheries Collection.*

Specimen Collections

Whilst all this was happening, another force impoverished the Galápagos, in the short-term at least: science. Towards the end of the nineteenth century, several scientific expeditions—mostly from Britain and the United States—charted a course for the archipelago to collect its plants and animals for zoos and museum collections. In 1905, the California Academy of Sciences carried out the last and by far the largest efforts to catalogue the natural productions of the islands. Their total haul of more than 75,000 specimens (including 264 giant tortoises) was greater than all previous expeditions combined. By today's standards, this might seem like a shocking statistic, but this was over one hundred years ago. If it appears as though these men had no conservation mindset, it's because they didn't, and nor did anybody else at the time. But they were concerned about extinction, which they considered a very real possibility for several of the unique species, especially the giant tortoises. The purpose of the expedition, according to the California Academy of Science's herpetologist Joseph Slevin, was 'to make an exhaustive survey, most extensive collections, and, most of all, to make a thorough study of the status of the gigantic land tortoises before it proved too late'.

In this goal, they succeeded and these collections—as epic as they were—have turned out to be of vital importance to the conservation movement that began to emerge in the 1950s. Without the historical background provided by the collections of the California Academy of Sciences and others, it would have been hard to think sensibly about conservation in the Galápagos.

Chapter 9. Humans: Part II

The importance we place on anniversaries can give them a transformative role, turning them into catalytic instants that can set us trundling along a genuinely novel path to the future. In the case of the Galápagos, Darwin-based anniversaries have played a major role in shaping the modern identity of the islands. As one historian has written, 'The Galapagos did not make Darwin; if anything, Darwin, through his superior abilities as a thinker and theoretician, made the Galapagos; and, in doing so, he elevated these islands to the legendary status they have today.'

The first Darwin anniversary of any note was in 1909 and marked what would have been the naturalist's one hundredth birthday (he'd died in 1882 at the age of seventy-three) and the fiftieth anniversary of the publication of *On the Origin of Species*. In comparison to Darwin-based celebrations that were to come, this was a fairly low-key, quintessentially

English affair; yet it was important for splicing Darwin and the Galápagos together in the popular consciousness.

The American naturalist William Beebe helped to cement this link in the 1920s with his immensely popular *Galápagos: World's End*. For historian Edward J. Larson, Beebe's published accounts of his 1923 expedition to the Galápagos 'espoused evolutionism of a near religious variety'. In Beebe's view, Darwin's account of the Galápagos had not been equalled for its 'general grasp and sheer interest'. With his beautiful writing style and knack for storytelling, Beebe pulled off something similar for the twentieth century, but his stunning descriptions of the Galápagos wilderness were also responsible for attracting a bunch of peculiar characters to the islands. Most infamous amongst these new arrivals were a German doctor with Nietzschean romantic leanings, his submissive patient-cum-lover and a sex-mad, self-styled 'baroness'. Within five years, the baroness had mysteriously vanished (was she murdered?), and the doctor was dead (was it food poisoning or something more sinister?).

The following year, 1935, offered a chance to draw a line beneath this intrigue and reaffirm the association between Darwin and the Galápagos: it was exactly one hundred years since the arrival of HMS *Beagle*. American travel writer Victor Von Hagen went to the Galápagos to mark the occasion. In the hold of his schooner, he carried a replica of the Darwin bust then on show at the American Museum of Natural History in New York. The plan was to erect it at the site of Darwin's first Galápagos footfall on San Cristóbal. As Von Hagen wrote, 'Raising the monument was more than an act of biological piety. It was the beginning of a campaign to bring to the attention of naturalists all over the world, and to the attention of the Republic of Ecuador . . . the need for conserving the irreplaceable natural phenomena of the archipelago, and to save from extinction this living laboratory for the study of evolutionary processes.'

In fact, Ecuador had already made moves to protect the Galápagos with a constitutional amendment in 1934 that placed several islands and most of the Galápagos fauna under protection. But the rest of the world was becoming increasingly vocal about the need to protect the archipelago.

We can be pretty certain, I think, that without the Darwinian connection, this concern wouldn't have been anywhere near as great.

The next big Darwin anniversary in 1959 provided the necessary impetus to do something with the islands. On 4 July that year, Ecuador passed an emergency law to coincide with the centenary of the publication of *On the Origin of Species*, formally declaring the uninhabited 97 percent of the Galápagos land mass the country's first national park. It was the first indication that it might yet be possible to realise Franklin D. Roosevelt's dying wish to preserve the Galápagos 'for all time as a kind of international park'.

Over in Europe, with the backing of UNESCO and its conservation-minded offshoot, the International Union for the Conservation of Nature (IUCN), a bunch of dedicated scientists simultaneously founded the Charles Darwin Foundation for the Galápagos Islands (CDF), an international, non-governmental organisation that aimed 'to provide knowledge and assistance through scientific research and complementary action to ensure the conservation of the environment and biodiversity in the Galápagos Archipelago'. For UNESCO's first director general, the visionary Julian Huxley, 'the preservation of all sources of pure wonder and delight, like fine scenery, wild animals in freedom, or unspoiled nature' was a key human need. Within five years of its foundation, CDF had an operative base, the Charles Darwin Research Station at the eastern edge of the then single-strip settlement of Puerto Ayora.

At the same time, a group of dedicated Americans kept up the momentum with the Galápagos International Science Project, a symposium attended by dozens of scientists, many of whom took the opportunity to get stuck into the geology and natural history of the islands. In the late 1960s, Ecuador's central government got around to appointing its first rangers to implement CDF's recommendations. Since then, the Galápagos National Park Service (GNPS) has steadily grown in stature, ultimately overshadowing CDF to become the single most important institution in the archipelago, employing more than two hundred staff with an annual operating budget of around $15 million. Over the course of the twentieth century then, science and conservation became central to the identity of the islands.

The Battle for the Tortoises

The GNPS has taken on plenty of battles. It's won some and lost some. There are too many to list—and it would be tedious to do so—but it's worth mentioning a couple of examples. These show what people can achieve when they put their minds to it and also reveal that there are some things that even the most dedicated conservationists cannot realistically hope to accomplish.

There are few success stories greater than that offered by the islands' giant tortoises. When CDF began to conduct the first-ever population surveys of the different tortoise species, they found them in a terrible way. Owing to the preference to go 'turpining' on those islands with a more accessible, easier-going terrain, the populations on Floreana, Santa Fé and Pinta had collapsed and were either extinct or effectively so. For these species, there was no escape.

The tortoises on Pinzón and Española would certainly have suffered the same fate had the conservation movement not come to the rescue. In the 1960s, Pinzón still had adequate numbers of adult tortoises, but they were old—very old. The invasion of egg- and hatchling-snaffling rats meant that no young tortoises were coming through the ranks, the last recruit to the Pinzón population probably hatching out some time in the nineteenth century. The CDF staff came up with the ingenious solution of excavating tortoise nests before the rats could get at them, then transporting eggs to Santa Cruz to be incubated, hatched and reared in captivity. When the tortoises were big enough to fend off a rat, the plan was to send them back to Pinzón. Almost fifty years later, this strategy is still in place, with the latest crop of captive-reared Pinzón offspring on show at the Charles Darwin Research Station on Santa Cruz.

On Española, the tortoises were in an even more perilous position. Like Floreana, Santa Fé and Pinta, Española is a relatively accessible island and had also been a favoured spot to harvest tortoises. A few remained, but so few that reproduction had come to a grinding halt. The presence of goats on the island only made matters worse. So conservationists rounded up the last fourteen Española tortoises and brought them to the research station on Santa Cruz. This small band of survivors (supplemented in 1975 by a rather randy Española male that had been hanging out at the San Diego Zoo) has been coaxed into producing over 2,000 offspring. In the meantime, the GNPS managed to eradicate goats from Española, paving the way for the reintroduction of these captive-born youngsters.

The detailed genetic description of each tortoise species has also proved to be very useful for conservationists. Nowhere is this more evident than on Isabela's Wolf Volcano. Here, geneticists have discovered hybrid tortoises with a wonderful diversity of genes, probably as a result of the activities of buccaneers and whalers moving these reptiles around in the not-so-distant past. Some of these hybrids show clear signs of recent San Cristóbal and Española ancestry. It also turns out that there are descendants of Floreana tortoises, a species that nobody has seen for more than 150 years. Several tortoises even have a smattering of Pinta ancestry. These are of special interest because the last-known Pinta tortoise, Lonesome George, died in 2012 after forty years in forlorn captivity. This makes it possible to think about restoring lineages assumed to have been lost long ago.

Project Isabela

On the larger, higher islands of Santa Cruz, Santiago and Isabela, the tortoises were able to find refuge from the tortoise hunters, but not from the invasion of mammals like pigs and goats. This demanded an extermination programme on an unprecedented scale, the outline of which was put together at a workshop held in the Galápagos in 1997. Project Isabela—an $8.5-million initiative that would synthesise and then build on the existing methods of dealing with alien mammals—was born.

The project started on Pinta, a small goat-infested island shaped like a blunt, spearhead just 11 km from top to bottom and 7 km from side to side. In previous decades, the GNPS had waged a protracted war against these destructive mammals, shooting more than 40,000 over the course of the 1970s alone. But after each goat blitz, it would emerge that a handful of individuals had somehow escaped the hunters' sights. Pinta became something of a workshop, a testing ground on which to hone eradication methodology, notably the deployment of so-called Judas goats.

Here's how it works. A goat is captured, fitted with a radio collar and released. As a member of a gregarious species, its natural instinct is to locate and then hang out with others of its kind, leading hunters to the last remaining individuals of a population. Although the idea had been around for several decades prior to Project Isabela, it was on Pinta in the Galápagos that the Judas goat became a serious conservation tool. Trials revealed that the most effective Judas goats were males that had been rendered sterile (with a nifty surgical snip rather than relatively crude castration). In addition, sterilised females treated with a cocktail of drugs to simulate a state of permanent oestrus proved to be an effective add-on, a supplementary tool now referred to as the Mata Hari goat (after the infamous World War I double agent).

With the successful eradication of Pinta's goats, Project Isabela began to scale up its operation, using a combination of carefully positioned fences to corral animals, aerial sharpshooting from helicopters, more conventional ground-based hunting, specially trained dogs and, where appropriate, the lethal deployment of Judas and Mata Hari goats to exterminate somewhere in the region of 200,000 large mammals from several islands, notably Santiago and northern Isabela. In the aftermath of Project Isabela, the GNPS took on feral goat populations on the inhabited islands. As of 2011, when the GNPS and its collaborators last revealed the status of these operations, there were no feral goats left in the Galápagos except for small populations on San Cristóbal, Santa Cruz and southern Isabela. Even there, however, the recovery of native vegetation has been impressive.

Of course, the work doesn't stop with a successful eradication. It's a rather dismal fact, but there are plenty of people who perceive that the conservation movement in the Galápagos has a negative influence on their livelihoods and are prepared to use goats as a political tool, threatening to reintroduce the animals to goat-free islands unless their demands are met. Between 2000 and 2010, it's reckoned that there were around ten incidents of such sabotage, on average one every year. These reintroductions are more trouble than they sound, as the effort required to seek out and destroy just a few individuals can be huge. In 2009, some malcontent set six goats down on Santiago, which had by then been a goat-free zone for some three years. The GNPS put the cost of removing these animals at $32,393. That's more than $5,000 a goat.

FIGURE 9.1. *The threat of sabotage is ever-present.* In 2004, local fishermen threatened to introduce goats to Fernandina, the most pristine island in the archipelago. *Reproduced from Victor Carrion et al., 'Archipelago-wide Island Restoration in the Galápagos Islands: Reducing Costs of Invasive Mammal Eradication Programs and Reinvasion Risk,'* PLoS One 6 (2011): e18835.

Project Pinzón

With the success of Project Isabela, the GNPS and its partners are now setting their sights on other mammalian invaders. Rats pose a serious problem, notably for invertebrates like the bulimulid snails, birds like the endangered medium tree finch, mangrove finch and Galápagos petrel, and reptiles like the Pinzón tortoise. But how do you get rid of rodents, which are not only fecund but can hide away in the most inaccessible of crevices? At present, the only approach with any chance of success is to put out poison.

At a workshop in the Galápagos in 2007 (precisely analogous to the one that gave rise to Project Isabela a decade earlier), conservationists laid the groundwork for what became known as Project Pinzón. As with Project Isabela, this began on small islands, incorporating lessons learned from each eradication attempt, revising the plan and gradually scaling up

FIGURE 9.2. **Project Pinzón.** Conservationists from Island Conservation (a nongovernmental organization specializing in the eradication of invasive species from islands) plan the distribution of rodenticide on South Plaza in 2012. © Rory Stansbury, Island Conservation.

to take on larger and larger islands. It's a massive, military-style undertaking that has been years in the planning. It has been necessary to carry out surveys of each island in detail prior to the release of any poison, to customize bait to suit the arid Galápagos environment, and to conduct a thorough assessment of the risks of the anticoagulant toxin to any of the many non-target species (including feeding the bait to Galápagos tortoises, which mercifully appear unaffected). For those species that face a risk of inadvertent poisoning, it was necessary to develop mitigation strategies. For the Galápagos hawk, it was thought best to bring all birds into captivity during the period they might conceivably dine out on poisoned rat, something that had never been achieved without the death of the hawk.

Finally, in January 2011, Project Pinzón entered its operational phase. A helicopter fitted with an industrial-scale hopper sent a steady spray of poison raining over several islets. At 4.9 km², the largest of these was Rábida, which was declared free of rodents in late 2012. With this success, the GNPS tackled South Plaza (0.2 km²) and Pinzón (18.2 km²) in November 2012. With three vessels acting as floating accommodation for a team of some forty-five staff from several collaborating institutions, the helicopter carried out an intensive double dosing of Pinzón. In 2013, with the island cleared of rats, baby tortoises began to emerge from the dusty Pinzón soil; these youngsters could be the first tortoises to hatch out on their island and survive into adulthood in more than one hundred years.

The plan is to move on to eradicate rodents from even bigger islands like Floreana (172 km²). If Floreana can be rendered rat-free, it should help the recovery of the Floreana mockingbird, shortly to be reintroduced to the main island from the two satellite islets on which it has survived. It might then be the moment to consider rodent eradication on Santiago. At 585 km², the size of this island will pose a significant challenge. That's not all. Although the Galápagos is notable for its paucity of land-based mammals, it seems that some kind of rodent reached the archipelago long before humans, giving rise to at least a dozen endemic species, collectively known as the Galápagos rice rats. Unfortunately, the introduction of bigger rats and predatory cats since the arrival of humans in 1535 is thought to have resulted in the extinction of most of these endemic rodents.

For most of the twentieth century, the Santiago rice rat (known only from a single specimen collected in 1906) was assumed to have suffered this irreversible fate. But in 1997, an American biologist and a couple of his students landed on Santiago's north shore, set up camp and put out a few traps to see what was there. The next morning the trio was astonished to find twenty-five Santiago rice rats in the traps. Somehow these tenacious little natives still survive alongside their more aggressive black rat cousins, perhaps owing to the abundance of food on this part of the island (in the shape of prickly pear opuntias) or possibly the rice rat's ability to tolerate long periods of drought. Whatever the reason, it is likely that a significant number of Santiago rice rats would have to be trapped and held in captivity prior to (and for several weeks after) the distribution of poisoned bait over the island. It's a complication for sure, but the Galápagos National Park Service has found workaround solutions to such challenges in the past.

This kind of restoration requires millions of dollars and decades of dedicated work, but this is something the Galápagos has made its own. This effort and other bold initiatives have made this archipelago a world leader in ecological restoration, a feat of which it can be rightly proud.

Insect Trouble

When it comes to alien invertebrates, the story is not quite so inspiring. There are certainly a lot of them—one in every six species in fact. There are several reasons for this. As a general rule, invertebrates are far less conspicuous than vertebrates. It's also true that most invertebrates are poorly understood. Where do they go after reaching Galápagos shores? How do they occupy their time? What impact do they have on native species? In most cases, the answers to such questions simply don't exist. If you don't have a good grasp of these fundamental details, it's tricky to come up with a method to control intruding populations. Even if you do, the complex life cycles of insects (with phases involving winged dispersal, for instance) stack the odds against success.

The little fire ant is a case in point. In 2000, the World Conservation Union's Invasive Species Specialist Group published a list of the one hundred worst invasive offenders, and this vicious ant is on it. In the Galápagos, it has been observed to displace native ants and spiders and may even affect the nesting success of birds and reptiles. 'On the larger islands the little fire ant is now distributed over thousands of hectares and is beyond the means of current methods of control,' wrote entomologists in 2005. Its successful eradication from the small island of Santa Fé in the 1990s, however, led to an attempt to eradicate it from a small (0.2 km^2) area of Marchena (a medium-sized island in the north of the archipelago). Throughout 2001, conservationists spread insecticide over the affected habitat and followed up with intense monitoring, using hot dogs and peanut butter to lure out any survivors. Over the course of six separate visits, they set up more than 160,000 bait stations in the affected area (nobody can accuse them of not being thorough) and found no sign of little fire ants at all. Not one. It looked like the insecticide had done its trick, leading the conservationists to suggest the effort was 'on the edge of success'. It probably was, but if you don't get every last one, all that effort counts for nothing. When a new infestation turned up in 2008, the most likely explanation was that some ants had survived the 2001 poisoning. The battle against the little fire ants on Marchena is ongoing and might yet be won. It will, however, be harder to take on invasive invertebrates like these on islands where they are more widely distributed.

Quinine and Other Demons

Efforts to eradicate invasive plants are even more troubled. Over the last few decades there have been serious efforts to kill off the worst offenders. Take the red quinine tree, for instance. It's thought that someone planted the first seed in 1946. Nobody really knows why. Maybe this person thought there might be some commerce to be had from exploiting its medicinal properties (quinine is known to disrupt the life cycle of the malaria parasite). Whatever the reason, it was unfortunate for many of the Galápagos' native plants. As a single tree can produce thousands of incredibly light seeds and the resultant seedlings can handle life in the shade, they were quick to establish, blazing a path through agricultural land, infesting the *Scalesia* forest and spreading into even more rarefied zones dominated by native *Miconia* and sedges (from around 500m above sea level and up). On Santa Cruz, the quinine tree is now thought to occupy more than 110 km², which corresponds to roughly 10 percent of the island.

Are these trees doing any harm to the rare habitats they've reached? Recent research suggests they are. The closer to one of these quinine trees, the fewer species there are and the thinner their ground coverage. This is particularly evident at the highest elevations, where the quinine cuts out almost all of the light on which low-lying grass, sedge, moss and fern rely. Although no native species appear to have gone extinct as a result, the invasion of this one plant has completely transformed the vegetation structure of these regions.

Rangers have experimented with several methods of eradication, starting by rooting out seedlings and saplings and ending up with the 'hack-and-squirt' method, attacking the trunk with a machete and then dressing the wounds with a lick of herbicidal poison. But it's expensive. It might cost more than $1 million to make one hack-and-squirt pass through all quinine-infested areas of Santa Cruz. If the goal is total eradication, it would take around $8 million in order to keep this up, year in year, year out, for at least a decade or two, and maybe longer.

Without a clear knowledge of how long seeds of a particular species can lie dormant in the earth, predicting such a time frame is little more than guesswork. There's also no guarantee of success. Of thirty such projects undertaken by the Galápagos National Park Service since the mid-1990s, only four have achieved their stated aim of eradication (and each of these four targeted species had a narrow range and was yet to get out of control). So there's the very real prospect—terrifying for those funding such an operation—that the species might not be successfully eradicated. Like the little fire ants on Marchena, the red quinine could just reappear. Then what would you have to show for your $8 million?

If eradication is not a realistic option, as it probably isn't for most introduced plants in the Galápagos, what's the alternative? The unpalatable truth is that we might have to live with them. Thankfully, many of the introduced species will not cause much damage, and we shouldn't lose too much sleep about their presence. Others will need to be controlled in hope of containing the impact they have rather than striving to get rid of them completely. In contrast to eradication, which has a clear end point, there is no end to control. As the Red Queen puts it in Lewis Carroll's *Through the Looking-Glass*, 'It takes all the running you can do, to keep in the same place.'

This is obviously going to be expensive, unless you can get another species to do the control for you. The parable of the cottony cushion scale insect is a case in point. This pest was first spotted in the Galápagos some thirty years ago, and by 2000 it had spread to most of the major islands in the archipelago. It's a problem because it sinks its mouthparts into the woody stems of plants, tapping into the sap of a range of species from the white mangrove to *Scalesia*. Interestingly, the presence of little fire ants may have furthered the scale insect's influence. The ants have a taste for the sugary honeydew secretion the scale insects produce during feeding, so they will carry these relatively immobile food factories with them when they move and may even defend them against predators. In spite of the minder role played by these ants, conservationists have had some success in keeping the scale insect in check, deliberately introducing yet another non-native insect (a ladybird) to the archipelago in 2002, a conservation manoeuvre unprecedented in the Galápagos. A decade on, the ladybird has become established and appears to be containing the scale insect's impact on native plants without itself causing any obvious untoward damage.

This kind of biological control is probably the only way to fight against the latest big deal, a parasitic botfly that goes by the scientific name *Philornis downsi*. In 1997, researchers working on Darwin's finches on Santa Cruz noticed something ugly, a woodpecker finch nest wriggling with some kind of maggot. The blood-filled guts of these larvae and their presence in the nostrils of a couple of the nestlings made it pretty clear what they'd been up to: feeding. If there were any doubt, one of the chicks soon died, and a post-mortem dissection uncovered a hole bored into the nestling's brain. The larvae of *P. downsi* are not fussy in their tastes. This parasite has probably been in the islands for a lot longer than just fifteen years, and we now know it is affecting at least fourteen species of land birds in the Galápagos (which include nine species of Darwin's finches). It is pretty well established, located on most islands. Several studies reveal the devastating, often lethal effect of these insects on nestlings.

As if this weren't enough, microscopic pathogens are a constant threat. The introduction of avian malaria and avian pox to Hawaii, for instance, was a major driver in the extinction of many of its native birds. In recent years, two mosquitos have been introduced to the Galápagos, including one that is the known vector for avian malaria. Other agents of disease, including avian influenza and West Nile virus, await the opportunity to skip over from the South American continent.

Due to concern over threats like these, conservationists helped draft the Special Law for Galápagos in 1998 (about which we'll hear more soon). This led to the creation of the Galápagos Inspection and Quarantine System (SICGAL) (amongst many other things), with the aim of preventing further introductions. Yet the challenge facing the SICGAL inspectors has been of epic proportions. It's not just the three commercial airlines now serving the Galápagos with more than 160 flights a month. Several cargo ships zip back and forth between Guayaquil and the islands, servicing them with fresh fruit and vegetables, drinking water, grain, beer, construction materials, furniture, fertilizers, vehicles, tyres, gas cylinders and so on. Concentrating on imports of the sixteen most common fresh food items in 2011 (including potatoes, bananas, plantains, yucca and onions), the cargo fleet brought an average of eight hundred tons of these goods to the Galápagos every month. Less than 2 percent of this produce was inspected upon departure or arrival.

How did we get to this point? How did the Galápagos become transformed from an inhospitable outpost where entrepreneurs had a tendency to be murdered to one with such a thriving economy? The answer is complex, of course, but several key influences deserve special mention. During World War II, the United States built a military base and airstrip on Baltra to protect the Panama Canal from a German or Japanese attack. It was a simple enough step, but one with a rather significant fallout for the archipelago. The presence of several thousand servicemen on Baltra demonstrated that, if adequately provisioned, the Galápagos could support a sizeable population. More people also meant greater economic prosperity for the early settlers. 'The Americans purchased all that could be fished, caught and produced. And at a good price,' wrote Norwegian Stein Hoff. With the Americans providing detailed oceanographic charts, the fishing became 'better than ever and the income sky high'. Most importantly, though, the airstrip made it possible to entertain commercial flights to the Galápagos and fuelled the rapid demographic expansion of Santa Cruz. Ironically, the Darwin legend, the protection afforded by conservationists and UNESCO's celebration of the Galápagos as its very first World Heritage Site in 1978 conspired with predictable consequence. People began to want to visit.

The Galápagos National Park Service responded by setting up its first dedicated visitor sites in the 1970s, short trails through the landscape marked with white posts from which tourists were not permitted to stray. It also instigated its guiding system, with tourists accompanied at all times by a highly qualified guide, someone who could act as a stand-in policeman to ensure nobody turned a blind eye to the rules. 'Take only pictures, leave only footprints' sums up the attitude. In many ways, it was a model operation. Beyond the little white posts and the weathered path, there is very little evidence that the nearly constant human footfall has done much to change the wider ecology or even the behaviour of the animals.

But the immediate impact of tourism on the landscape—which is what the Galápagos authorities paid most attention to at first—is not really the issue. With the benefit of hindsight, it seems obvious that those making decisions about the future of the Galápagos should have paid more attention to human behaviour. The international community, the Ecuadorian government, and the Galápagos authorities should have anticipated the population expansion that occurred during the twentieth century and taken serious steps to minimise the damage it would cause. Even conservationists failed to build a human dimension into their thinking, and it's easy to see why. For most conservation-minded folk, their love of the natural world got them into the business. But we now know—and the Galápagos is a perfect case in point—that if conservationists fail to consider the needs of humans, they ignore a vital part of the ecology.

Chapter 10. Humans: Part III

In 1950, there were just over 1,000 people living in the Galápagos. By 1990, there'd been roughly a tenfold increase to almost 10,000. Up until this point, any Ecuadorian national—from Azogues to Zamora—was free to move to the Galápagos, just as one might decide to move from London to Manchester or from New York to San Francisco. Non-Ecuadorians found it easy to get the necessary paperwork too. By 2000, the population had reached nearly 20,000.

New Build

Where did all these people go? Very few tourists will venture beyond the main seafront strip of Puerto Ayora, but a ten-minute walk to the north and east of the town's touristic hub gives the answer. One arrives in a part of town called La Cascada, named after a waterfall that tumbles off the nearby cliff when it rains. This neighbourhood, however, is far from the beatific idyll its name suggests. From the mid-1990s onwards, newcomers were able to purchase plots from the Santa Cruz municipality and build on them as they saw fit. As a consequence, La Cascada is a rather schizoid neighbourhood, where it's possible to find a well-constructed two-story home butting up against a flimsy lean-to with a corrugated roof. The streets are impossibly narrow, many with no pavement and barely wide enough to accommodate a pickup truck, let alone a fire engine. Only after buildings went up did the municipality get around to thinking about electricity, water supply and how to deal with sewage. Needless to say, planning for this kind of infrastructure should take place before the first brick is laid. Doing it the other way round is quite patently ridiculous, not to mention expensive, and has effectively guaranteed the residents of this neighbourhood a set of second-rate municipal services prone to dysfunction. This kind of backward development also explains some bizarre statistics that fall out of Ecuador's national census, like the fact that the proportion of Galápagos homes serviced with a networked sewer did not increase between 2001 and 2010. It fell from 30.8 to 26.8 percent. La Cascada is clear evidence of an absence of long-term thinking on the part of the Santa Cruz municipality.

During the 1990s, there was increasing tension between elements of the human population and the conservation community. Things came to a head in 1995 when a group of disgruntled fishermen bore down on the offices of the Galápagos National Park Service (GNPS), unhappy about quotas the conservationists were imposing on the extraction of sea cucumbers and hence their livelihoods. There were rowdy stand-offs, Molotov cocktails and death threats, notably against then-GNPS director Arturo Izurieta. The unsightly display of violence filled column inches of local, national and international print and was captured on radio and television. This kind of unrest rumbled on for another decade, with conservationists and fishermen coming head-to-head each season until the eventual collapse of the sea cucumber fishery.

Special Law

This kind of conflict was getting nobody anywhere. In 1998, Ecuador passed the Special Law for Galápagos, an explicit acknowledgement of a simple fact: what goes for the other twenty-three provinces in Ecuador might not be appropriate in this extraordinary archipelago. The special law is a long document. We've already heard of one of its achievements (the creation of the Galápagos Inspection and Quarantine System, or SICGAL). Here are a few more: it saw that all the revenue from the entrance fee to Galápagos National Park remained in the islands rather than being siphoned off to the central government in Quito (as had been the case); it marked the formal creation of the Galápagos Marine Reserve (all 133,000 km^2 of it); there was also an effort to resolve the conflict over the marine reserve's resources, bringing all those with a stake in them (including representatives of the fishing, tourism and conservation sectors) around the same table; perhaps most importantly, the special law introduced measures to contain domestic immigration. After 1998, it was no longer possible for just anyone with an Ecuadorian passport to relocate to the Galápagos, and it became harder for non-Ecuadorians to obtain the necessary permits. There were now conditions.

Yet, for all its good intentions, the Special Law for Galápagos proved impossible to implement. Ecuador was in economic meltdown, triggered in part by the collapse of world oil prices (its main export) in the 1980s. The national currency, the sucre, tumbled in value and was eventually abandoned altogether in 2000 in favour of the American dollar. As inflation soared out of control, presidents came and went. In the decade from 1996 to 2006, Ecuador had eight different leaders. There was mass unemployment, unrest and emigration, with more than one in six citizens fleeing in search of a more predictable future elsewhere.

The political chaos rippled its way down the chain of command, eventually washing up on the rocky shores of the Galápagos. Between 1998 and 2006, there were six different ministers of the environment, six provincial governors of the Galápagos and eight directors of the Galápagos National Institute (INGALA), a body created in 1980 to oversee the human side of the Galápagos. At the GNPS, the situation was worse still, with more than ten directors and acting directors over the same period. There is no revolving door to the offices of the GNPS. But if there had been, it wouldn't have stopped spinning.

It was not just lack of leadership. The 1980s witnessed 'the bureaucratisation of the Galápagos', with just about every department of Ecuador's central government replicating itself—in miniature—in the islands. By 2006, there were more than fifty central government organisations operating in the Galápagos, representing everything from fishing, tourism and the environment to health, education and welfare, not to mention dozens of national and international nongovernmental organizations. With so many voices, all clamouring for attention and all with a different stake in the islands, it's not that surprising that the vision for the Galápagos as enshrined in the special law got lost along the way.

United Nations

All of this was of considerable concern to UNESCO's World Heritage Committee, which sent a crack team of assessors to the islands for a week-long, fact-finding mission in early 2006. They did not like what they saw. Their report runs to almost fifty pages and gives a stark sense that things were out of control. Here's a view of the Galápagos (as the UNESCO assessors saw it) in 2006.

In spite of the regulations on domestic migration introduced by the special law, the Galápagos population continued to expand at almost 7 percent a year, making it the fastest-growing province in Ecuador. One in five of these residents was an irregular or illegal migrant without the necessary paperwork. In the 1960s, there had been just two commercial flights a week; now there were more than thirty. Visitor numbers were increasing at an incredible 12 percent per year. There was a risk of oil spills, an issue that came to the fore when the tanker *Jessica* ran aground in 2001, releasing 285,000 gallons of fuel into the Galapagos Marine Reserve. A cruise ship carrying around five hundred passengers had been granted a permit to operate in Galápagos waters. The Inspection and Quarantine System (SICGAL) was struggling; from 2002 to 2010, the number of inspectors decreased by 25 percent (from forty to thirty), whereas imports increased by 60 percent (from around 35,000 to more than 55,000 tons). There was a rapid increase in the number of hotels; in the fifteen years from 1991 to 2006, the number of places to stay more than doubled from twenty-six to sixty-five. A new airport went up on Isabela without appropriate consultation. The prospect of sports fishing loomed. Political infighting was rampant. Conclusion: 'Galápagos is shifting into an economic development model that is fundamentally at odds with long term conservation and sustainable development interests.'

UNESCO distilled this report into a list of fifteen recommendations and called for a multi-stakeholder meeting to be held by March 2007 at the very latest. In the end, this took place in the Galápagos in April 2007. The UNESCO representatives noted that none of the fifteen specific issues from 2006 been acted on. 'On the contrary,' they reported, 'there were clear indications that the situation was getting worse.' It reads a little like a school report.

If Ecuador's newly incumbent president Rafael Correa (who came to power in January 2007) felt patronised, he didn't show it. He responded immediately by issuing a bold statement of intent towards the Galápagos. On 10 April, in the middle of the meeting, he and elite members of his cabinet signed a decree declaring the archipelago 'in a state of risk' and a 'national priority'. In view of this, the UNESCO World Heritage Committee decided to stick the Galápagos on its list of properties 'in danger', citing the combined threats of invasive species, unbridled tourism, immigration and overfishing. It was, they said, a measure intended 'to draw attention to its state of conservation and mobilise international assistance'. In the opinion of leading conservationists in the Galápagos, Correa's call to action in 2007 offered 'the local, national, and international communities what might be the last opportunity to implement a strategic change in direction in Galápagos'. It drew a stark line in the black volcanic sand.

In 2007, INGALA began efforts to deport illegal migrants from the Galápagos and to tighten its control on immigration. The GNPS put a stop to the arrival of more big cruise vessels, a clear acknowledgement that some tourism activities are more suitable than others. It also began a shake-up of the way that Galápagos tourism operates, starting work on a new, self-sustaining tourism model for the islands. A new constitution, passed in 2008, made Ecuador the first country in the world to grant rights to nature—if nothing else, an important symbol of intent.

In recognition of these steps, the World Heritage Committee agreed to take the Galápagos off the list of properties 'in danger' in 2009, though many conservationists felt this was premature. The unbridled growth of tourism in the islands goes to the heart of the matter. It's been estimated that around 10 percent of all international visitors to Ecuador visit the Galápagos, but they are of far greater value to the national economy,

bringing hundreds of millions of dollars into the country, which is equivalent to more than half the gross national income from tourism.

This extraordinary fact explains why there has never been a cap on
the number of tourists visiting the Galápagos. The influx of tourists to
the archipelago is responsible for the kind of economic growth that most
politicians can only dream of. Between 1999 and 2005, the Galápagos had one of the fastest-growing economies anywhere in the world,
with the total income coming into the islands increasing by 78 percent
(equivalent to an average annual growth rate of almost 10 percent). No
surprise then that unemployment in the Galápagos should be far lower
than in mainland Ecuador. Not only that but the employed majority is
able to earn, on average, more than three times what they could on the
continent. Other indicators, like the treatment of women, levels of education, health care and access to the Internet, all suggest that Galápagos
tourism has significantly raised the standard of living in the archipelago.
Since the implementation of the Special Law for Galápagos, the visitor
fee (currently $100 if you're not Ecuadorian and $6 if you are) has also
been put to good use, being distributed among several key institutions,
including the GNPS (to the tune of 45 percent).

There is a flip side of course. With the Galápagos promising jobs,
personal wealth and a relatively high standard of living, emigration from
the mainland has continued apace. Though the Special Law for Galápagos sought to contain this human traffic, there are several ways to get
to live and work in the Galápagos. If you can demonstrate that you are
bringing skills that cannot be met by the existing labour force, you've
got a good chance of securing temporary residency. With few permanent
residents interested in cement mixers and breeze blocks, for instance,
the construction industry is overwhelmingly made up of this migrant,
manual workforce. Another way of staying within the letter of the special law (though not its spirit) is to marry a permanent resident. It's hard
to know how common this is, but it is happening. In 2007, there were
thirty-eight marriages registered in the Galápagos. In 2010, that number
had shot up to 232. Over the same period—2007 to 2010—the number
of divorces rose from precisely none to sixty-four. If a sham marriage
seems a bit much, it's always possible just to turn up on a tourist permit
(which gives you ninety days in the archipelago), then hide. There are

plenty of employers prepared to turn a blind eye to a glaring absence of paperwork. Others take a more clandestine route into the Galápagos, so their name doesn't enter the system at all.

Whatever the route, whatever the balance between permanent, temporary and illegal residents, the rapid influx of people to the Galápagos over the past two decades means that a large proportion are recent migrants (74 percent according to the Ecuadorian National Census of 2010). There are at least a couple of reasons why this might be of concern. First, there is a feeling that recent migrants don't appreciate the challenges of island living like lifelong islanders, but come with mainland mindsets that end up changing Galápagos culture. To emphasise this point, a 2009 survey of over 1,000 households across the four inhabited islands found that more than half identified with an 'expansionist' mentality, showing 'a strong motivation for development, through mainland and island transportation, tourism, and construction'. Second, rapid immigration has radically altered the age structure of Galápagos society, with the twenty-something bracket notably distended. So, even if there were no more domestic immigration, period, the population is still likely to grow as all those biological clocks count down to zero.

Into this painful social transition walked Raquel Molina, a biologist by training who became the director of the GNPS in 2006. Apart from her gender, which some in the still patriarchal Galápagos society may have had issues with, Molina stood out for her uncompromising attitude towards environmental protection. She was, one might imagine, just the kind of tough-talking, no-nonsense leader that the Galápagos was crying out for. But the way in which her term of office unfolded gives pause for thought.

Shortly after her appointment, Molina set the tone for her tenure in an interview published in Ecuador's *El Comercio* newspaper, accusing a wealthy and influential local businessman of fishing illegally in the Galápagos Marine Reserve and levelling death threats at one of her staff. A few months later, in March 2007, just prior to the multi-stakeholder meeting, she and several colleagues were assaulted by members of the Ecuadorian navy whilst trying to clamp down on an illegal tourism operation on Baltra. In spite of a couple of months in hospital, Molina waded into even deeper political water the following year, confronting

Ecuador's minister for the environment (technically her boss) over a permit for a tourist vessel. Marcela Aguiñaga pulled rank and fired the insubordinate Molina, citing her inability to 'adapt to the daily routines of the job and the ministry's priorities'.

A memo from the US consulate in Guayaquil back to Washington, DC, offers a rather less opaque explanation for the sacking. 'Molina's true crime was not incompetence; instead, it was her unwillingness to turn a blind eye to corruption and unlawful activities in the park,' wrote Consul General Douglas Griffiths in his confidential report, which was published by WikiLeaks in 2008. According to Griffiths, Metropolitan Touring (one of Ecuador's largest tour operators) had entered into an agreement with a couple of Galápagos vessels that would see their sixteen-passenger permits transferred to *La Pinta* (Metropolitan's newly refurbished luxury yacht). Molina declined to sanction the step, citing regulations that prohibit a single ship bundling up multiple permits. In an effort to bypass the obstructive Molina, Metropolitan appealed to the higher authority of Minister Aguiñaga. In her former career as a lawyer representing Galápagos tour operators, Aguiñaga had counted Metropolitan as one of her clients, and she ordered Molina to sign off on *La Pinta*'s permits. 'Everything's been arranged in Quito,' an employee of Metropolitan Touring crooned to the US consulate. When the GNPS director stood firm, Aguiñaga—asserting the legality of the permit—decided to call time on Molina's career. 'As a result of her unwillingness to compromise her principles, she had many enemies amongst those looking to exploit the economic opportunities the islands offer,' wrote Consul General Griffiths. *La Pinta* is now a registered tour vessel with a permit to operate at its full forty-eight-berth capacity.

It's hard to know what to make of this. Maybe it's accurate; maybe it's just one version of events. Whatever the truth, there is no denying the widespread perception that a new species is now endemic in the Galápagos: corruption. 'Despite clear legal limits on the number of fishing licenses, boat permits and resident visas issued in the Galápagos, corruption has allowed many to skirt the regulations, and resulting environmental consequences have been terrible,' concluded Griffiths.

Quite clearly, the Galápagos is no longer just a biological laboratory; it is now a social, political and economic laboratory too. The results of experiments conducted in other island labs, such as the Seychelles, the Mascarenes, Hawaii and New Zealand, have rarely turned out well for the native flora and fauna. Yet, in spite of all that's happened, the Galápagos is still one of the least-touched, best-preserved natural wonders in the world. This is partly because humans came late to the islands, only showing a real interest once the whaling industry got going in the early nineteenth century. It is also because the archipelago has been blessed by an incredibly powerful brand. The link with Charles Darwin gave the Galápagos a certain symbolic appeal in the minds of the pioneering conservationists of the 1950s. Come 1959 and the centenary of the publication of *On the Origin of Species*, Ecuador embraced the cause, marking out 97 percent of the Galápagos land mass as national park, thereby confining any human population to just 3 percent.

It is understandable that much of the popular news coverage of the Galápagos should zero in on what humans are doing to the islands. It is this, after all, that poses the greatest threat to the future of many of the archipelago's unique species. Yet the only reason this is news at all is because there is still so much to marvel at in the Galápagos, still so much to lose. At this moment it seems appropriate to ask, What if? What if the international community had not kept a spotlight on the islands? What if Ecuador had not afforded this far-flung territory any protection at all? How would it look today?

In some senses, not so different. There would be international tourism. There would be a significant resident population. There would be a thriving local economy. But the negative consequences of human occupation would probably be far greater. It's likely that we would have taken over just about every inch of the species-rich highland habitat with the inevitable extinction of thousands of species. We would have occupied most of the major islands. There would be more roads, more cars, more traffic. All manner of alien species would have taken root or gained a foothold to the detriment of the native ecology. Looking to the horizon, there would be a steady stream of cruise liners and huge fishing vessels struggling to reap the last of the archipelago's rich marine resources.

As it is, there are still pockets of relatively intact highland habitat to marvel at. Humans occupy only four of the islands. With the 30,000-strong population concentrated in just two towns (Puerto

Baquerizo Moreno and Puerto Ayora), the most damaging consequences of human settlement are confined to just two islands (San Cristóbal and Santa Cruz). On the other two occupied islands, Floreana and Isabela, the resident population is sufficiently small that it's been possible to contain, even reverse, some of the damage that humans have caused. On many islands, invasive donkeys, pigs, goats and rats have been removed. There is a moratorium on cruise liners. Fishing is tightly regulated. In short, the situation in the Galápagos could be so much worse.

This is not a reason for complacency, but it's very important to acknowledge. To do so is to recognise something so obvious it is often overlooked: the decisions we have made in the past have guided the Galápagos into a very different, more isolated, pristine present than the one that might have been. For those who care about the future of the Galápagos, this should be an uplifting admission. With a long-term vision for the archipelago, it is surely possible to protect much of what makes it unique. Islands like Floreana, which is the focus of a concerted restoration initiative, are even likely to recover some of their ecological innocence. On islands like San Cristóbal and Santa Cruz, where this may not be possible owing to the heavy human footprint, a lot can still be done to slow the attrition.

Since humans first set foot in the Galápagos almost five hundred years ago, these islands have had a considerable impact on human thought. In the words of Robert Bowman, an ornithologist who carried out a survey in the islands on behalf of UNESCO in 1957, 'No area on Earth of comparable size has inspired more fundamental changes in Man's perspective of himself and his environment.' We still have a lot to learn from the Galápagos. With the global population placing an ever-greater burden on the world's natural resources, there is an urgent need to come up with better ways of managing the conflict between humankind and nature. In the Galápagos, this has been the explicit goal for more than fifty years. There have been failings. There have been triumphs. We can learn from both. The future that Ecuador and the international community are forging for the Galápagos is by no means certain. It will be fascinating to see how it unfolds.

Acknowledgements

A big thank you to everyone at Profile Books for all their support with *The Galápagos*, particularly the late Peter Carson, Andrew Franklin, Penny Daniel, Rebecca Gray, Anne-Marie Fitzgerald, Cecily Gayford, Sarah Hull and Michael Bhaskar. Thanks to George Lucas of InkWell Management and to all the attentive folks at Basic Books, notably Thomas Kelleher, Sandra Beris, Tisse Takagi and Elizabeth Dana.

I'd like to express my gratitude to all institutions and individuals that contribute to www.archive.org. This is a tremendous resource, giving access to so many old and often rare books, which takes a lot of the slog out of writing a book like this. There are two books on the Galápagos that I have found a particular inspiration: *Evolutions's Workshop* by Edward J. Larson is a brilliant account of the history of science in the Galápagos, and *Darwin in the Galápagos*, by Thalia Grant and Greg Estes re-creates Darwin's movements through the archipelago in unbelievable detail, exploring these islands' importance to his thinking. I am also indebted to John Woram for creating Las Encantadas (http://www.galapagos .to), an absolutely fabulous website full of Galápagos-related texts and images, to John van Wyhe and colleagues for Darwin Online (http://darwin-online.org.uk), a super resource for leafing through Darwin's published and unpublished writings, and to Jim Secord and colleagues at the Darwin Correspondence Project (http: //www.darwinproject.ac.uk). For more recent scientific journals, I relied heavily on Ben Norman of John Wiley & Sons, who pinged over dozens of PDFs on all sorts of obscure subjects. Apart from reading a lot, I have benefitted from the wisdom of many people, including Randal Keynes, Swen Lorenz, Ole Hamann, Alan Tye, Fritz Trillmich, Gabriele Gentile, Patricia Parker, Karl Campbell, Julia Pooder, Matthew James, Gisella Caccone, James Gibbs, Stephen Blake, Fausto Llerena, Joe Flanagan, Patricia Jaramillo, Washington Tapia, Charlotte Causton, Graham Watkins, Mark Gardener, Samantha Singer, Peter Quintanilla, Marilyn Cruz, Hugo Echeverria, Luis Die, Desiree Cruz, Santiago Bejarano, Rachel Dex, Richard Montagu and Jemma Pearson. In particular, I'd like to express my extreme gratitude to Thalia Grant, Gregory Estes, Tim Birkhead, Linda Cayot, Felipe Cruz, Rashid Cruz, Dennis Geist, Conley McMullen, Christine Parent, Marta Tufet and Mark Wilson for taking the time to read excerpts, sample chapters and, in several cases, a draft of the entire book. In spite of all this help, any errors contained between these covers are obviously my own.

In my capacity as editor of *Galápagos Matters*, I am fortunate to have the support of all those at the Galápagos Conservation Trust in the United Kingdom, including Ian Dunn, Jen Jones, Pete Haskell, Victoria Creyton, Isabel Banks and Leah Meads, not to mention former colleagues Leonor Stjepic, Toni Darton, Abigail Rowley, Kate Green and Nicholas Moll. Over at the Galápagos Conservancy in the United States, it is a pleasure to work with Johannah Barry, Linda Cayot, Lori Ulrich and Rebecca Fuhrken. Lori Ulrich contacted several members of the Galápagos Conservancy who kindly gave permission to reproduce the most wonderful photographs in the colour section. I'd like to give a special mention to all those at Island Conservation (especially Heath Packard, Karl Campbell, Erin Hagen, Brad Keitt and Nick Holmes) for helping me research a feature on rat eradication I wrote for Helen Pearson at *Nature* (some of which I've worked into this book).

Thank you to all my friends and family, especially my parents John and Stella, Tom and Ana, Mary and Mark, Hugh and Sheila, Zaid and Kate, Matt and Marisa, John and Sara, and the Celeriac XI. But it's really Charlotte who deserves the greatest thanks. Without her love and support, this book simply wouldn't exist.

If you would like to find out more, visit my blog at the *Guardian* (http://www.theguardian.com/science/animal-magic), show that you 'like' the book on Facebook (https://www.facebook.com/TheGalapagosByHenryNicholls), or follow me on Twitter (@WayOfThePanda).

Appendix A: How to Visit the Galápagos

I first visited the Galápagos as a tourist in 2003. In the years that followed, I turned down invitations to lecture on board cruise ships and resisted the urge to return. My logic: 'I've been fortunate enough to go to the Galápagos once,' I kept telling myself. 'It is now for others to have their turn.'

In 2011, I got the chance to explore this kind of thinking with Magaly Oviedo, then head of tourism at the Galápagos National Park Service (GNPS). 'There are lots of people who come to the Galápagos I never want to see again,' she said. 'But people who really appreciate the value of the Galápagos and understand the challenges of preserving it, they can come as many times as they like.' I liked her answer. The lives of thousands of people depend on Galápagos tourism, so it's just plain silly to demonise it out of hand. But we can always do better. In writing this book, I have come to appreciate the Galápagos more than ever and reached a more profound understanding of the challenges of preserving it. I am not so bold as to imagine this book will alter the footfall on the Galápagos, but I do hope it might make it just a little lighter.

Several exciting developments have a direct bearing on how we will see the Galápagos in years to come. As we've learned, the Special Law for Galápagos of 1998 established a mechanism for the distribution of the park entrance fee to local institutions. Unfortunately, some bright spark of a lawyer actually spelt out the fee structure—in figures—thereby making any increase illegal. This explains why it has remained fixed for almost fifteen years. At the time of writing—in September 2013—an amendment to the special law is currently being considered by the Ecuadorian congress, which would see several important changes to the way the entrance fee is levied and distributed.

If this lengthy amendment is signed off on in its entirety, the all-new special law would see the entrance fee tied to 'the basic unified wage of the general worker', nimbly keeping pace with a fundamental economic touchstone in Ecuador. The fee structure would also begin operating on a sliding scale to encourage visitors to stay longer in the islands. If you spent less than three nights, for instance, you would feel it with a $300 hit to your wallet. If your visit were three to six nights, you'd only pay $187. If you were to stay for seven or more, it'd just be $160. Ecuadorian nationals would no longer get in for just $6 but would have to pay the same as everyone else. Finally—and crucially—the amendment to the special law would see the GNPS receive a much larger proportion of the entrance fee (up from 45 percent, as is currently the case, to 70 percent).

There will inevitably be those who grumble that it is to become more expensive to visit the Galápagos, and some, presumably, will feel this puts it beyond budget. But this ignores the simple fact that visitors have been being undercharged for years. Go for seven or more nights and pay just $160. Frankly, it's a bargain. If this doesn't persuade you, consider the wider effects of the fee structure. The sliding scale should discourage weekend breaks and encourage visitors to spend more time (and hence money) in the islands. It will also generate a more substantial and more secure income for all the local institutions that benefit from the fee. In 2010, for instance, 173,296 visitors generated $11,541,644 in fees. Although 45 percent of this—$5,193,739—went to the GNPS, this barely covered a third of the institution's annual expenditure of $14,112,340. The rest was funded by the central government, fines and the sale of permits to tour operators. Had the new fee structure been in place, however, it would have generated $21,318,072 from the same number of visitors, of which some $15 million (70 percent) would have gone to the GNPS, making it the first self-sufficient national park in the world.

What about the other institutions that currently receive a cut of the entrance fee? As the proposed changes to the special law currently stand, the remaining 30 percent would go to the municipalities (25 percent) and the parish councils (5 percent). The other bodies that currently receive a slice of the entrance fee would instead be supported by funds direct from the central government. At present, these institutions obviously have a vested interest in seeing rapid growth in Galápagos tourism: they will get more money. By taking a large proportion of the humans in the Galápagos out of this monetary loop, the case for capping tourist numbers becomes one that—from a political perspective at least—will be easier to make. It's very realistic that there will be a cap on numbers, Magaly Oviedo told me back in 2011.

There has also been a shake-up in the itineraries that the tourist vessels take around the islands. In the past, the distribution of tourists across the seventy terrestrial and seventy-five or so marine visitor sites in the Galápagos was far from even. The popularity of Punta Suarez on Española, for instance, meant that it was frequently operating beyond its carrying capacity. Since 2012, tourist vessels have been working to a new set of guidelines, one that prevents vessels visiting any one site more than once in any two-week period.

If you are only planning to go for a week, this may have a bearing on what vessel you decide to travel on. If, for instance, you want to look into a caldera, you'll have to book a tour that will take you to Sierra Negra on Isabela. If it's the highlands that you're interested in seeing, make sure your itinerary takes you into the heart of either San Cristóbal, Santa Cruz or Floreana. If you just have to see a waved albatross, you'll have to head to the Galápagos between April and December and book yourself on a vessel that will take you to Punta Suarez on Española. If you have a burning desire to swim with lots of sharks, you must choose a cruise that will take in the outlying islands of Darwin and Wolf.

If you don't have a wish list of sites or tick list of species, don't feel the need to invent one. Every visitor site in the Galápagos will have something to marvel at. You will not be disappointed. During the course of a typical week-long Galápagos cruise, you will see plenty of geological formations and a wide variety of plant life; you will have ample opportunity to peek beneath the waves, gape at pelicans, boobies, and frigatebirds, and step over marine iguanas and get close to tortoises.

If you are prepared to adopt this more laissez-faire attitude to your itinerary, then you need to ask whom you want to show you the islands' riches. Since 1975, the GNPS has been training naturalist guides to chaperone visitors around the islands, gently policing their behaviour and giving them an incredibly informative experience into the bargain. Inevitably, however, the quality of guide can vary, from class I (the bottom-end guide) to class III (the gold standard). If you end up with someone whose language skills are poor, whose grasp of the natural history isn't quite up to scratch and whose heart just isn't in it, you are bound to be disappointed. If, however, you insist on a Class III guide, you stand a much better chance of being satisfied.

The vessel itself might be something that affects your decisions. In 2011, there were ninety-three vessels with tourist permits, ranging from the smallest (with just eight berths) to the largest (with one hundred berths). There is little to distinguish among them in terms of mod-cons (most of them operate to extremely high standards), but for many people, the size of vessel is a relevant consideration. Small vessels offer a more intimate, personal experience with the obvious drawback that you might end up with people you don't get on with. A larger vessel, by contrast, gives you more opportunity to seek out kindred spirits (and greater space to distance yourself from those with whom you're not so chummy). But if it's the wilderness experience you seek, a large vessel may not be what you're after, with each shore expedition requiring several tenders to ferry more passengers into your immediate horizon. For many visitors, the type of vessel can matter too, some preferring a motorboat to a yacht or a catamaran to a single-hulled vessel. It's horses for courses.

Another consideration is who owns the vessel you are thinking of travelling on. The smaller ships tend to be owned by Galápagos residents and employ a larger proportion of Galápagos-based staff. The larger ships, by contrast, tend to be owned by national or international tour operators. With a smaller vessel, your money will take a fairly direct and transparent course into the Galápagos. The bigger and more international the operation, the harder it is to know where your money ends up, passing through a more complex web of middlemen. But many of these larger tour operators are in a position to demonstrate their corporate responsibility, making serious contributions to the work of the Charles Darwin Foundation and the GNPS. It's worth asking around. In addition, the larger the vessel, the more efficient it tends to be, an economy of scale helping to give passengers a rather tidier ecological footprint. 'It makes sense to have a mix,' said Oviedo. If, in the future, the GNPS introduces a green certification scheme for tourist vessels, this will certainly help visitors make more informed decisions when arranging their trip.

Whichever vessel you settle for, I can virtually guarantee that, with a quality guide, you will not be disappointed. By travelling from island to island, usually from one day to the next, you may even get a sense of the subtle variation in species that Darwin seems to have sensed. But if sailing is not your thing, there are other ways to visit the Galápagos.

Ever since the 1960s, there has been a steady—and, more recently, explosive—rise in the number of conventional, land-based hotels. In 1982, there were just eighteen hotels in the Galápagos, catering to a maximum of 214 guests. By 2006, this had risen to sixty-five, with accommodation for 1,668. These hotels provide an important source of income for a significant proportion of the Galápagos population with little or no access to the cruise dollars that remain largely offshore. But all this construction—much of it unregulated and unplanned—has resulted in a mishmash of hotels from the frankly substandard to the undeniably luxurious. Since 2007, there has been a moratorium on further construction of hotels, but if you intend to stay in one of those with a license to operate, be sure you know what you're letting yourself in for.

For the land-based tourist, the four urban centres are all served by one or more visitor sites. But if you want to see more of the Galápagos, you'll need to make other plans. A few sites might be accessible by road, with local operators or a surfeit of taxis prepared to take you out of town. Most visitor sites, however, are only accessible by sea, so booking onto a small vessel that operates day tours is your best bet. It's preferable to staying in your hotel, but be warned. These trips will not take you very far—usually to a visitor site a short way along the coastline or to a nearby islet—and they can leave you feeling a little rushed.

One advantage of staying in a hotel rather than on a cruise ship is that you'll get a much better sense of the human side of the Galápagos. It might not be as attractive as a blazing white beach covered in basking sea lions, as cute as a couple of pint-sized Galápagos penguins or as moving as a mockingbird pecking at your shoelaces, but taking just a couple of hours to wander beyond the main street of Puerto Ayora can still be surprising.

If you really want to get involved in the human side of the Galápagos, you might consider volunteering for the GNPS (galapagospark.org) or one of several charities working in the islands, such as the Charles Darwin Foundation (darwinfoundation.org), Galápagos ICE (galapagosice.org), Fundación Bolivar Education (ecuadorvolunteers.org) or A Broader View Volunteers (abroaderview.org). It won't be the conventional Galápagos experience, but it'll certainly leave you with a far more profound understanding of the intractably complex future facing the islands. As a bonus, the GNPS will waive the entrance fee to the islands for most of these schemes (though it remains uncertain whether this will still be the case when it moves to the more expensive sliding scale outlined above).

Finally, become a Galápagos evangelist. Join up with your local Friends of Galápagos Organization (Appendix B), which will help you stay in touch with the goings-on in the archipelago (even if you don't return), support the conservation of the islands and remember how fortunate you were to see these remarkable islands and all their inhabitants.

Whatever you do, do not take the Galápagos for granted.

Appendix B: Friends of the Galápagos

Finland
Nordic Friends of Galápagos
www.galapagosnordic.org

Germany
Zoologische Gesellsschaft Frankfurt
www.zgf.de

Japan
The Japanese Association for Galápagos
www.j-galapagos.org

The Netherlands
Stichting Vrienden van de Galápagos Eilanden
www.galapagos.nl

New Zealand
Friends of Galápagos New Zealand
www.galapagos.org.nz

Switzerland
Freunde der Galápagos Inseln
www.galapagos-ch.org

United Kingdom
The Galápagos Conservation Trust
www.savegalapagos.org

United States
The Galápagos Conservancy
www.galapagos.org

Appendix C: Maps and Diagrams

Several key phenomena are absolutely crucial for a proper understanding of the Galápagos. I hope these maps and diagrams help to illustrate the volcanoes, tectonic plates, hotspot, currents and habitats that have gone into making these extraordinary islands.

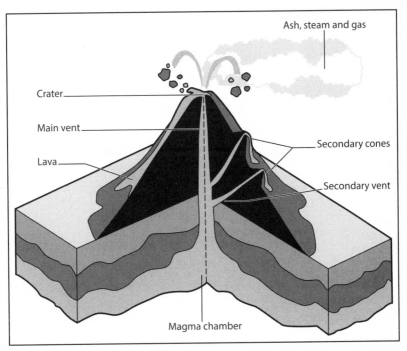

FIGURE C.I. Charles Darwin was intrigued by what went on inside a volcano and proposed the presence of an internal chamber of molten rock with the densest lava located near the bottom and the least dense rising to the top. This would allow lavas of different composition to emerge in one eruption and account for his observation that the rocks on the lower slopes of Santiago were darker and denser than those at the verdant summit.

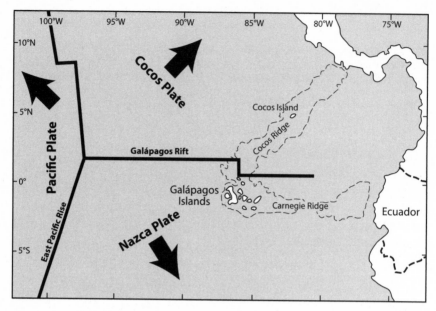

FIGURE C.2. The Galápagos sits near the junction of three major tectonic plates: the Pacific, the Cocos and the Nazca. At the faults between these plates, magma bubbles up from the earth's mantle, cooling to form new crust and driving the three plates in different directions. The East Pacific Rise runs from south to north, along the junction between the Pacific Plate to the west and the Cocos and Nazca Plates to the east. The Galápagos Rift is oriented from east to west between the Cocos Plate to the north and the Nazca Plate to the south. The islands that we call the Galápagos are the manifest pinnacles of a now submerged mountain range known as the Carnegie Ridge.

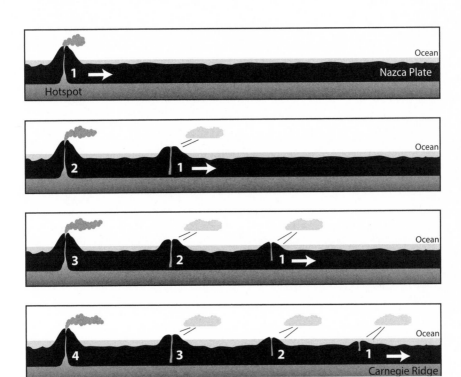

FIGURE C.3. The Galápagos Archipelago is explained by the existence of a deep-seated hotspot that periodically sends volcanoes bubbling to the surface of the Nazca Plate. Courtesy of the interaction between the three major tectonic plates, the Nazca Plate is moving steadily to the south-east and carrying each island—as if on a conveyor belt—away from the hotspot. As this happens, the island cools, shrinks and weathers away, eventually assuming a monumental position as a seamount on the undersea Carnegie Ridge.

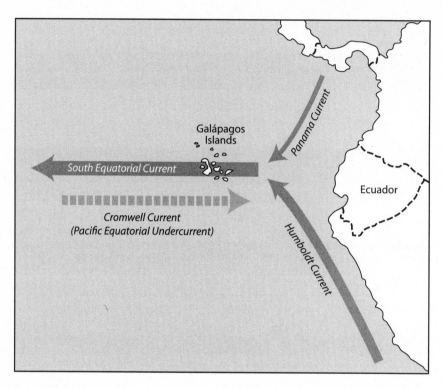

FIGURE C.4. The interplay between oceanic currents has a profound effect upon the Galápagos ecology and climate. The strong, cold Humboldt Current flows up the coast of Peru, combining with the weaker, warmer Panama Current from the north to form the South Equatorial Current, which surges out into the Pacific from east to west along the equator. The Cromwell Current (or Pacific Equatorial Undercurrent) runs in the opposite direction. Between June and November, the Humboldt Current is the major player, its cold, north-westerly influence turning down the temperature in the islands and causing moisture in the warmer air to condense into a drizzling mist, or garúa. When the Humboldt slackens in December, the Panama Current becomes ascendant, the mist evaporates and the hot season begins. As the temperature of the water rises, evaporation increases, clouds grow and rain falls.

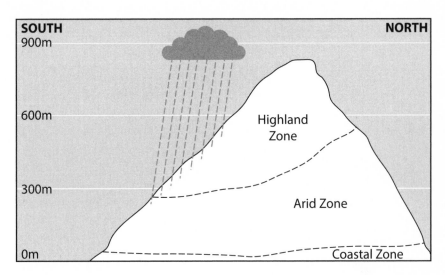

FIGURE C.5. The Galápagos climate accounts for the existence of three principal ecological zones: the coastal zone is characterized by its saltiness; the arid zone is particularly dry; and the highlands, which are fed by the *garúa* during the cool season and rain during the hot season, support an altogether more effulgent flora.

Notes

Epigraphs

vi **'The natural history of these islands':** Charles Darwin, *Journal of Researches into the Natural History and Geology of the Countries Visited During the Voyage of H.M.S. Beagle Round the World*, 2nd ed. (London: John Murray, 1845), 377.

vi **'Charles Darwin effected the greatest':** Julian Huxley, 'Charles Darwin: Galápagos and After,' in *The Galápagos: Proceedings of the Symposia of the Galápagos International Scientific Project*, edited by Robert Bowman (Berkeley: University of California Press, 1966), 3.

vi **'Only one English word':** Kurt Vonnegut, *Galápagos: A Novel* (New York: Delacorte Press, 1985).

vi **'No area on Earth of comparable size':** Robert Bowman, 'Contributions to Science from the Galápagos,' in *Galápagos: Key Environments*, edited by Roger Perry (Oxford: Pergamon Press, 1984), 278. American ornithologist Robert Bowman's vision helped forge the current identity of the Galápagos. In 1957, UNESCO sent Bowman and Austrian-born zoologist Irenäus Eibl-Eibesfeldt to conduct a survey of the islands. As a direct consequence of their report, the Galápagos National Park and the Charles Darwin Foundation came into being in 1959.

Prologue

ix **'the establishment of such military bases':** Paul Harrison, *Study of the U.S. Air Forces' Galápagos Islands Base* (1947), 1. Available at http://www .galapagos.to/TEXTS/USAF1947.HTM.

ix **The classified report identified two possible sites:** US Navy, *Field Monograph of Galápagos Islands* (US Navy Department of Naval Intelligence, 1942). See http://www.galapagos.to/MAPS/ONI-92.HTM.

x **'These Islands represent the oldest form of animal life':** Franklin D. Roosevelt to Secretary of State Cordell Hull, 'Memorandum for the Secretary of State,' 9 February 1944, Las Encantadas, http://www.galapagos .to/WW2/19440209.HTM.

x   **'I have been at this for six or seven years':** Franklin D. Roosevelt to Secretary of State Cordell Hull, 'Memorandum for the Secretary of State,' 1 April 1944, Las Encantadas, http://www.galapagos.to/WW2/19440401.HTM.

xi   **Since tourism to the islands began:** For an in-depth look at the Galápagos tourism industry and associated statistics, see Bruce Epler, *Tourism, the Economy, Population Growth, and Conservation in Galápagos* (Puerto Ayora: Charles Darwin Foundation, 2007). Available at http://www.galapagos.org/wp-content/uploads/2012/01/TourismReport.pdf.

xii   **Of those 4,000 native species:** In July 2012, there were 4,107 native species registered in the Charles Darwin Foundation's online database, of which 17 were extinct (e-mail from Dolma Alonso, 11 July 2012). See the foundation's online database at http://datazone.darwinfoundation.org for the latest numbers.

xiii   **'is like dross, worthless':** Letter from Fray Tomás de Berlanga, the bishop of Panama, to His Sacred Imperial Catholic Majesty King Charles I of Spain, 26 April 1535. See http://www.galapagos.to/TEXTS/BERLANGA.HTM.

xiv   **'wretched-looking little weeds':** Darwin, *Journal of Researches*, 374.

Chapter 1

1   **'It is as though the earth had spewed forth rocks':** Between 1935 and 1962, Eleanor Roosevelt wrote a regular column, titled 'My Day', that was syndicated across the United States. The quotations here come from her column of 4 April 1944. See http://is.gd/Usyniq.

2   **'Geology is a capital science':** Letter from Charles Darwin to William Fox, 1 July 1835, sent from Lima. See Darwin Correspondence Database, http://www.darwinproject.ac.uk/entry-282 (accessed 27 August 2013).

2   **'I look forward with joy':** Letter from Charles Darwin to John Henslow, 12 August 1835, sent from Lima. See Darwin Correspondence Database, http://www.darwinproject.ac.uk/entry-283 (accessed 27 August 2013).

2   **'A broken field of black basaltic lava':** Darwin, *Journal of Researches*, 373.

2   **'strange Cyclopean scene':** Ibid., 374.

4   **'In many places the coast is rock-bound':** Herman Melville, 'The Encantadas or, Enchanted Isles' (1854), published in *Putnam's Monthly Magazine*. This quotation comes from his first sketch of 'The Isles at Large'.

4   **On 23 February 1535, a galleon set sail:** All details of Tomás de Berlanga's trip to the Galápagos come from his letter to King Charles I of Spain.

4   **most likely Española:** See John Woram, *Darwin Slept Here* (New York: Rockville Press, Inc., 2005).

5   **The American explorer Benjamin Morrell:** All details of Benjamin Mor-
        rell's Galápagos adventure come from Benjamin Morrell, *A Narrative of
        Four Voyages to the South Sea, North and South Pacific Ocean, Chinese Sea,
        Ethiopic and Southern Atlantic Ocean, Indian and Antarctic Ocean, from the
        Year 1822 to 1831* (New York: J. and J. Harper, 1832), 191–195.

7   **'There is dire mischief':** Melville, from his fourth sketch, 'A Pisgah
        View from the Rock'.

7   **'high up, we saw a small jet':** Richard Keynes, ed., *Charles Darwin's
        Beagle Diary* (Cambridge: Cambridge University Press, 2001), 357,
        entry for 30 September 1835 (Darwin Online, http://is.gd/z1B4JT).

8   **'proof of elevation to a small degree':** Charles Darwin, *Geological Diary:
        Galapagos* (1835), CUL-DAR37, 716-795A, transcribed by Thalia Grant
        and Gregory Estes (Darwin Online, http://is.gd/h2Kb2K). Darwin made
        this discovery at Cerro Tijeretas at the bottom of Frigatebird Hill on San
        Cristobal, one of the many sleuth-like deductions made by Grant and Estes
        in their forensic study of Darwin's movements through the Galapagos. See
        Grant and Estes, 'Darwin in the Galapagos: His Footsteps Through the
        Archipelago,' *Notes and Records of the Royal Society* 54 (2000): 343-368;
        and Grant and Estes, *Darwin in the Galapagos: Footsteps to a New World*
        (Princeton, NJ: Princeton University Press, 2009).

9   **The eruption of Fernandina in 1968:** Tom Simkin and Keith Howard,
        'Caldera Collapse in the Galápagos Islands, 1968,' *Science* 169 (1970):
        429–437.

9   **'One night I slept on shore':** Darwin, *Journal of Researches*, 374.

10   **at Sullivan Bay on Santiago:** Sullivan Bay and Bartolomé are both
        named after Bartholomew Sulivan, one of HMS *Beagle*'s surveyors,
        though the unusual spelling of his surname is rarely replicated on mod-
        ern maps.

10   **'folds of drapery, cables and . . . the bark of trees':** Charles Darwin,
        *Geological Observations on the Volcanic Islands Visited During the Voyage of
        H.M.S. Beagle* (London: Smith Elder and Co., 1844), 105. See http://is.gd
        /tJb8Rw.

11   **'one of the rare uses of marmalade pots':** Tom Simkin, 'Geology of
        Galápagos,' *Biological Journal of the Linnean Society* 21 (1984): 61–75.

11   **'had their southern sides':** Darwin, *Journal of Researches*, 373.

11   **'this singular uniformity in the broken state':** Ibid.

11   **Darwin guessed that the amorphous mass:** Darwin, *Geological Observa-
        tions*, 101.

13   **'less regular fourth line':** To find out what Darwin wrote about 'fissures
        of eruption' see ibid., 116.

13   **'The principal craters':** Darwin, *Geological Observations*, 116.

14   **'In 1963, a Canadian geophysicist':** All J. Tuzo Wilson's quotations come from his groundbreaking paper on hotspots: J. Tuzo Wilson, 'A Possible Origin of the Hawaiian Islands,' *Canadian Journal of Physics* 41 (1963): 863–870.

14   **'a major fracture in the Earth's crust':** J. P. Eaton and K. J. Murata, 'How Volcanoes Grow,' *Science* 132 (1960): 925, cited in Wilson, 'A Possible Origin,' 1963.

15   **Of the volcanoes that do still peak:** For estimates of the ages of each island, see Vincent Neall and Steven Trewick, 'The Age and Origin of the Pacific Islands: A Geological Review,' *Philosophical Transactions of the Royal Society Series B: Biological Sciences* 363 (2008): 3293–3308.

15   **Studies that have taken a closer look:** Karen S. Harpp et al., 'The Cocos and Carnegie Aseismic Ridges: A Trace Element Record of Long-Term Plume-Spreading Center Interaction,' *Journal of Petrology* 46 (2005): 109–133.

15   **Other work has even found evidence:** D. M. Christie et al., 'Drowned Islands Downstream from the Galápagos Hotspot Imply Extended Speciation Time,' *Nature* 355 (1992): 246–248.

Chapter 2

17   **On 9 April, gung-ho American naturalist:** The reconstruction of William Beebe's pioneering Galápagos dives is based on William Beebe, *The Arcturus Adventure: An Account of the New York Zoological Society's First Oceanographic Expedition* (New York: Putnam, 1926). Unless otherwise indicated, all of Beebe's quotations come from this source.

20   **They were looking for signs:** See Peter Lonsdale, 'Clustering of Suspension-Feeding Macrobenthos Near Abyssal Hydrothermal Vents at Oceanic Spreading Centers,' *Deep Sea Research* 24 (1977): 857–863. This article contains the first grainy photographs of the ocean floor at the Galápagos Rift.

20   **In 1977, the US National Oceanic and Atmospheric Administration:** For details of the *Alvin* dive, see John B. Corliss et al., 'Submarine Thermal Springs on the Galápagos Rift,' *Science* 203 (1979): 1073–1083.

20   **The BBC's 2006 three-part documentary:** Richard Wollocombe, personal communication, 14 September 2006.

22   **'Four very conspicuous, strong, curved, canine teeth':** Leonard Jenyns, 'Part IV. Fish,' in *The Zoology of the Voyage of H.M.S. Beagle*, ed. Charles Darwin (London: Smith, Elder and Co., 1838–1843), 101.

22   **'loud grating noise':** Ibid., 154.

22   **'without exception':** Ibid., vi.

22   'Fish, Shark & Turtles': For this and other quotations from this fishing expedition, see Keynes, *Charles Darwin's* Beagle *Diary*, 353, entry for 17 September 1835.

22   The Beagle's captain, Robert FitzRoy: See Robert FitzRoy, *Narrative of the Surveying Voyages of His Majesty's Ships* Adventure *and* Beagle . . . (London: Henry Colburn, 1839), 505 (Darwin Online, http://is.gd /qLvoCE).

23   more than 3,500 mm of rain: Peter Grant, 'Extraordinary Rainfall During the El Niño Event of 1982–83,' *Noticias de Galápagos* 39 (1984): 10–11.

23   'stupendous thunderstorms': 'News from Academy Bay,' *Noticias de Galápagos* 38 (1983): 3.

23   In 1935, Norwegian settler Alf Kastdalen: Alf Kastdalen, 'Changes in the Biology of Santa Cruz Island Between 1935 and 1965,' *Noticias de Galápagos* 35 (1982): 7–12.

23   In 1983, Kastdalen: 'News from Academy Bay.'

24   'mottled brown': Jenyns, 'Fish,' 11.

25   It is thought that in the 1980s: Günther K. Reck, *The Coastal Fisheries in the Galápagos Islands, Ecuador* (PhD diss., Christian-Albrechts-Universität, 1983).

25   By the time of the first official census: Calculations of sea cucumber densities made from Alex Hearn, 'Sea Cucumber Fishery,' in *Galápagos: Preserving Darwin's Legacy*, ed. Tui De Roy (London: A&C Black, 2009), 87.

26   During the particularly harsh El Niño: Fritz Trillmich and D. Limberger, 'Drastic Effects of El Niño on Galápagos Pinnipeds,' *Oecologia* 67 (1985): 19–22.

26   Sea lions, by contrast: S. Villegas-Amtmann et al., 'Multiple Foraging Strategies in a Marine Apex Predator, the Galápagos Sea Lion *Zalophus wollebaeki*,' *Marine Ecology Progress Series* 363 (2008): 299–309.

26   In the 1982–1983 El Niño: Trillmich and Limberger, 'Drastic Effects.'

27   A bull sea lion simply can't fight: Ulrich Pörschmann et al., 'Male Reproductive Success and Its Behavioural Correlates in a Polygynous Mammal, the Galápagos Sea Lion (*Zalophus wollebaeki*),' *Molecular Ecology* 19 (2010): 2574–2586.

27   Unless the conditions are really good: Fritz Trillmich and Jochen Wolf, 'Parent-Offspring and Sibling Conflict in Galápagos Fur Seals and Sea Lions,' *Behavioural Ecology and Sociobiology* 62 (2008): 363–375.

27   Cruising to the west of the islands: All quotations from come from James Colnett, *A Voyage to the South Atlantic and Round Cape Horn into the Pacific Ocean . . .* (London: W. Bennett, 1798), 147.

28 **But the actions of Colnett and his fellow whalers:** Judith Denkinger et al., 'From Whaling to Whale Watching: Cetacean Presence and Species Diversity in the Galápagos Marine Reserve,' in *Science and Conservation in the Galápagos Islands*, ed. Stephen Walsh and Carlos Mena (Springer, 2012), 217–232.

28 **'We examined a large number of them':** Robert Snodgrass and Edmund Heller, 'Shore Fishes of the Revillagigedo, Clipperton, Cocos and Galápagos Islands,' *Proceedings of the Washington Academy of Sciences* 6 (1905): 333–427.

Chapter 3

32 **the gaudier the feet of the foster father:** Alberto Velando et al., 'Male Coloration and Chick Condition in Blue-Footed Booby: A Cross-Fostering Experiment,' *Behavioural Ecology and Sociobiology* 58 (2005): 175–180.

33 **Give him back his food:** Alberto Velando et al., 'Pigment-Based Skin Colour in the Blue-Footed Booby: An Honest Signal of Current Condition Used by Females to Adjust Reproductive Investment,' *Oecologia* 149 (2006): 535–542.

33 **Only males in fine fettle:** Roxana Torres and Alberto Velando, 'Male Reproductive Senescence: The Price of Immune-Induced Oxidative Damage on Sexual Attractiveness in the Blue-Footed Booby,' *Journal of Animal Ecology* 76 (2007): 1161–1168.

33 **The conclusion is clear:** Alberto Velando et al., 'Pigment-Based Skin Colour.'

34 **Dulled-down females received much less attention:** Roxana Torres and Alberto Velando, 'Male Preference for Female Foot Colour in the Socially Monogamous Blue-Footed Booby, *Sula nebouxii*,' *Animal Behaviour* 69 (2005): 59–65.

34 **Those ornithologists who've taken a good look:** David Anderson, 'The Role of Hatching Asynchrony in Siblicidal Brood Reduction of Two Booby Species,' *Behavioural Ecology and Sociobiology* 25 (1989): 363–368.

34 **juvenile females suffer higher mortality:** Terri Maness et al., 'Ontogenic Sex Ratio Variation in Nazca Boobies Ends in Male-Biased Adult Sex Ratio,' *Waterbirds* 30 (2007): 10–16.

35 **Occasionally, these advances are sexual too:** David Anderson et al., 'Non-breeding Nazca Boobies (*Sula granti*) Show Social and Sexual Interest in Chicks: Behavioural and Ecological Aspects,' *Behaviour* 141 (2004): 959–977.

35 **they concluded, 'provide the first evidence':** Martina Müller et al., 'Maltreated Nestlings Exhibit Correlated Maltreatment as Adults:

Evidence of a "Cycle of Violence" in Nazca Boobies (*Sula granti*),' *Auk* 128 (2011): 615–619.

35   **Indeed, the disappearance of hawks:** The red-footed booby colony on San Cristóbal is at Punta Pitt, a site you can visit at the eastern-most point on the island. The colony on Floreana is actually on the islet of Gardener-near-Floreana, which you can't visit as this is a highly protected site and one of the last strongholds of the Floreana mockingbird.

36   **'a curious communal game':** Friedmann Koster and Heide Koster, 'Twelve Days Among the "Vampire-Finches" of Wolf Island,' *Noticias de Galápagos* 38 (1983): 4–10.

36   **We now know that the brownness:** Patricia Baiao et al., 'The Genetic Basis of the Plumage Polymorphism in Red-Footed Boobies (*Sula sula*): A Melanocortin-1 Receptor (MC1R) Analysis,' *Journal of Heredity* 98 (2007): 287–292.

37   **'the brown makes them less susceptible':** Mattieu Le Corre, 'Plumage Polymorphism of Red-Footed Boobies (*Sula sula*) in the Western Indian Ocean: An Indicator of Biogeographic Isolation,' *Journal of Zoology* 249 (2006): 411–415.

37   **This suggests that perhaps these birds:** Carlos Valle et al., 'Plumage and Sexual Maturation in the Great Frigatebird *Fregata minor* in the Galápagos Islands,' *Marine Ornithology* 34 (2006): 51–59.

38   **These assaults can be brutal:** Charles Hedley, 'Hunting Trick of the Man-o'-War Hawk,' *Emu* 25 (1926): 279–281; D. Birch, 'The Feeding Ecology of Greater Frigatebirds *Fregata minor* and Lesser Frigatebirds *F. ariel* on Aride Island,' *Phelsuma* 8 (2000): 29–33.

38   **it looks like the extraordinary puffed-up courtship:** Donald Dearborn et al., 'Courtship Display by Great Frigatebirds, *Fregata minor*: An Ener-getically Costly Handicap Signal?' *Behavioural Ecology and Sociobiology* 58 (2005): 397–406.

39   **Darwin made this observation himself:** In the Galápagos, Darwin noted, 'nearly every land-bird, but only two out of the eleven marine birds, are peculiar'. Charles Darwin, *On the Origin of Species by Means of Natural Selection, or the Preservation of Favoured Races in the Struggle for Life* (London: John Murray, 1859), 390 (Darwin Online, http://is.gd/qK4Zcd).

39   **waved albatrosses have even colonised:** M. P. Harris, 'The Biology of the Waved Albatross *Diomedea irrorata* of Hood Island, Galápagos,' *Ibis* 115 (1973): 483–510.

40   **By feeding at night:** Jack Hailman, 'The Galápagos Swallow-Tailed Gull Is Nocturnal,' *Wilson Bulletin* 76 (1964): 347–354.

40   **They are also probably after particular prey species:** Sebastian Cruz et al. 'At-Sea Behavior Varies with Lunar Phase in a Nocturnal Pelagic Seabird, the Swallow-Tailed Gull,' *PLoS ONE* 8 (2013): e56889.

40   **The swallow-tailed gull's bright-orange eye ring:** Andrew Iwaniuk et al., 'Morphometrics of the Eyes and Orbits of the Nocturnal Swallow-Tailed Gull (*Creagrus furcatus*),' *Canadian Journal of Zoology* 88 (2010): 855–865.

40   **The parents will brood the young chick:** Harris, 'The Biology of the Waved Albatross,' 491–492.

40   **The signal from one bird:** David Anderson et al., 'At-Sea Distribution of Waved Albatrosses and the Galápagos Marine Reserve,' *Biological Conservation* 110 (2003): 367–373.

42   **Still, there's a lot of death:** David Anderson et al., 'Population Status of the Critically Endangered Waved Albatross *Phoebastria irrorata*, 1999 to 2007,' *Endangered Species Research* 5 (2008): 185–192.

42   **In the particularly harsh El Niño:** Hernan Vargas et al., 'Biological Effects of El Niño on the Galápagos Penguin,' *Biological Conservation* 127 (2006): 107–114; Carlos Valle and Malcolm Coulter, 'Present Status of the Flightless Cormorant, Galápagos Penguin and Greater Flamingo Populations in the Galápagos Islands, Ecuador, After the 1982–83 El Niño,' *Condor* 89 (1987): 276–281.

42   **The flightless cormorant, by contrast:** Martyn Kennedy, 'The Phylogenetic Position of the Galápagos Cormorant,' *Molecular Phylogenetics and Evolution* 53 (2009): 94–98.

42   **Whereas all other twenty-six species:** Rory Wilson, 'What Grounds Some Birds for Life? Movement and Diving in the Sexually Dimorphic Galápagos Cormorant,' *Ecological Monographs* 78 (2008): 633–652.

44   **'I believe that the nearly wingless condition':** Darwin, *On the Origin of Species*, 134.

44   **As with much else in the Galápagos:** Dee Boersma, *The Galápagos Penguin: A Study of Adaptations for Life in an Unpredictable Environment* (PhD diss., Ohio State University, 1974).

44   **It's the male that's left:** R. Tindle, 'The Evolution of Breeding Strategies in the Flightless Cormorant (*Nannopterum harrisi*) of the Galápagos,' *Biological Journal of the Linnean Society* 21 (1984): 157–164.

Chapter 4

48   **other plants like sea sandwort:** Sturla Fridriksson, 'Life Develops on Surtsey,' *Endeavour* 6 (1982): 100–107.

48   **'Amongst other things, I collected every plant':** Letter from Charles Darwin to John Stephen Henslow, [28–29] January 1836. See Darwin

Correspondence Database, http://www.darwinproject.ac.uk/entry-295 (accessed 31 August 2013).

48   **'It never occurred to me':** Charles Darwin, *Narrative of the Surveying Voyages of His Majesty's Ships* Adventure *and* Beagle . . . *Journal and Remarks. 1832–1836* (London: Henry Colburn, 1839), 474 (Darwin Online, http://is.gd/Zsc1ZM).

48   **'I do not want you to take any trouble':** Letter from Charles Darwin to John Stephen Henslow, 3 November 1838. See Darwin Correspondence Database, http://www.darwinproject.ac.uk/entry-429A (accessed 31 August 2013).

49   **Within days of receiving the Galápagos plants:** Letter from Joseph Hooker to Charles Darwin, [12 December 1843–11 January 1844]. See Darwin Correspondence Database, http://www.darwinproject.ac.uk/entry -723 (accessed 31 August 2013). The figure of 217 different plant species comes from the table Darwin included in the second edition of his *Journal* published immediately after this correspondence in 1845.

49   **Of these so-called endemic species:** Joseph Hooker, 'On the Vegetation of the Galápagos Archipelago, as Compared with That of Some Other Tropical Islands and of the Continent of America,' *Transactions of the Linnean Society of London* 20 (1847): 235–262.

49   **This fact, wrote Hooker, 'quite overturns':** Letter from Hooker to Darwin, 1843–1844. In his *Principles of Geology*, Charles Lyell had argued that regions of high endemicity were clearly '*centres* or *foci* of creation'. But since most of Darwin's Galápagos plants appeared to belong each to its own island, the idea of a single focus of creation in the Galápagos was clearly suspect.

49   **'I cannot tell you how delighted':** Letter from Charles Darwin to Joseph Hooker, [11–12 July 1845]. See Darwin Correspondence Database, http://www.darwinproject.ac.uk/entry-889 (accessed 31 August 2013).

49   **'Reviewing the facts here given':** Darwin, *Journal of Researches*, 398.

49   **Hooker mapped out several possible routes:** Hooker, 'On the Vegetation of the Galápagos,' 253.

49   **'On several parts of the shore':** Colnett, *A Voyage to the South Atlantic*, 58.

50   ***Beagle* captain Robert FitzRoy:** FitzRoy, *Narrative*, 505.

50   **Although Hooker observed that many:** Hooker, 'On the Vegetation of the Galápagos,' 253.

50   **'It is quite surprising that the Radishes':** Letter from Charles Darwin to Joseph Hooker, 19 April [1855]. See Darwin Correspondence Database, http://www.darwinproject.ac.uk/entry-1669 (accessed 31 August 2013).

50   **'the seeds of 14/100 plants':** Darwin, *On the Origin of Species*, 359.

51   **Darwin fed a pigeon on seeds:** Ibid., 361.

51   **This helps explain why the vast majority:** Conley McMullen, 'Breeding Systems of Selected Galápagos Islands Angiosperms,' *American Journal of Botany* 74 (1987): 1694–1705.

52   **Owing to 'the excessive minuteness':** Hooker, 'On the Vegetation of the Galápagos,' 257.

52   **'Living birds can hardly fail':** Darwin, *On the Origin of Species*, 361.

52   **Although some Galápagos plants clearly:** Duncan Porter, 'Geography and Dispersal of Galápagos Islands Vascular Plants,' *Nature* 264 (1976): 745–746.

52   **On Santiago, Darwin stuck a thermometer:** Keynes, *Charles Darwin's Beagle Diary*, 363. 'The sand was intensely hot, the Thermometer placed in a *brown* kind immediately rose to 137, & how much higher it would have done I do not know: for it was not graduated above this' (entry for 16 October 1835).

53   **'such wretched-looking little weeds':** Darwin, *Journal of Researches*, 374.

53   **'On most of the isles':** Melville, 'The Encantadas.' These quotations come from his first sketch, 'The Isles at Large'.

54   **there are six different species:** Conley McMullen, *Flowering Plants of the Galápagos* (Ithaca, NY: Cornell University Press, 1999).

54   **Darwin collected just one:** John Henslow, 'Description of Two New Species of *Opuntia*,' *Magazine of Zoology and Botany* 1 (1837): 466–468.

54   **'strongly resembling hog's bristles':** Ibid.

56   **the palo santo tree:** The palo santo tree (*Bursera graveolens*), which translates from the Spanish as 'holy stick', gets its name from its aromatic, much prized resin that is similar to frankincense and myrrh.

57   **'The wood gradually becomes greener':** Keynes, *Charles Darwin's Beagle Diary*, 355, entry for 25 September 1835.

59   **In spite of considerable differences:** Edward Schilling et al., 'Evidence from Chloroplast DNA Restriction Site Analysis on the Relationships of *Scalesia* (Asteraceae: Heliantheae),' *American Journal of Botany* 81 (1994): 248–254.

Chapter 5

61   **'Alas! it ejected some intensely acrid fluid':** Francis Darwin, *The Life and Letters of Charles Darwin, Including an Autobiographical Chapter* (London: John Murray, 1887), 1:50.

62   'never collected in so poor a country': Darwin, *Narrative*, 473.

62   But the Galápagos sits well below this line: Stewart Peck, *Smaller Orders of Insects of the Galápagos Islands, Ecuador: Evolution, Ecology, and Diversity* (Ottawa: NRC Research Press, 2001), 22.

62   In the Galápagos, there are just ten: Ibid.

62   The intensity of the equatorial sun: It may also be that the Galápagos carpenter bee is a relatively recent arrival and there has not been sufficient time for divergence to occur.

62   'Even when an occasional large': William Beebe, *Galápagos: World's End* (New York: G. P. Putnam's Sons, 1924), 102.

64   He watched in amazement as thousands: Darwin, *Journal of Researches*, 160.

64   In 1992, in the midst of an El Niño: Peck, *Smaller Orders of Insects*, 48.

64   A couple of ornithologists, camping out: Koster and Koster, 'Twelve Days.'

64   Some 735 of 1,555 native insects: Peck, *Smaller Orders of Insects*, 25.

65   The Galápagos flightless weevils: Andrea Sequeira et al., 'Nuclear and Mitochondrial Sequences Confirm Complex Colonization Patterns and Clear Species Boundaries for Flightless Weevils in the Galápagos Archipelago,' *Philosophical Transactions of the Royal Society Series B: Biological Sciences* 363 (2008): 3439–3451.

65   looking at terrestrial invertebrates as a whole: Of 2,198 native species, 1,174 are endemic. See Peck, *Smaller Orders of Insects*, 27.

65   the bulimulid land snails: Bulimulidae is a taxonomic family of air-breathing land snail found in the Americas. So the group is not endemic to the Galápagos, though all Galápagos bulimulids are endemic.

65   'It occurred to me that land-shells': Darwin, *On the Origin of Species*, 397.

65   Perhaps there was another way: Ibid.

66   'a duck or heron might fly': Ibid., 385.

66   In a recent study, researchers fed land snails: Shinichiro Wada, 'Snails Can Survive Passage Through a Bird's Digestive System,' *Journal of Biogeography* 39 (2012): 69–73.

67   Snails trying to make it at different altitudes: Christine Parent, 'Diversification on Islands: Bulimulid Land Snails of Galápagos' (PhD diss., Simon Fraser University, 2008).

67   which appears to be around 40m above sea level: Christine Parent, personal communication, 22 January 2013.

Chapter 6

71   **'Amongst the species of this family':** Nora Barlow, ed., 'Darwin's Ornithological Notes,' *Bulletin of the British Museum (Natural History): Historical Series* 2 (1963): 261 (Darwin Online, http://is.gd/XPOByb). Historian Frank Sulloway calculates that Darwin penned his ornithological notes between 18 June and 19 July 1836; see Frank Sulloway, 'Darwin's Conversion: The *Beagle* Voyage and Its Aftermath,' *Journal of the History of Biology* 15 (1982): 325–396.

72   **'a gradation in form of the bill':** Barlow, *Darwin's Ornithological Notes*, 261.

72   **Darwin's thirty-one finch specimens belonged:** In fact, Darwin only came across nine species of Galápagos finch. Of these, he managed to identify six species; see Frank Sulloway, 'Darwin and His Finches: The Evolution of a Legend,' *Journal of the History of Biology* 15 (1982): 1–53.

72   **'Seeing this gradation and diversity':** Darwin, *Journal of Researches*, 380.

74   **'I fortunately happened to observe':** John Gould (1841), 'Part 3. Birds,' in Darwin, *The Zoology of the Voyage*, 63.

74   **'each variety is constant in its own Island':** Richard Keynes, ed., *Charles Darwin's Zoology Notes and Specimen Lists from H.M.S. Beagle* (Cambridge: Cambridge University Press, 2000), 298.

74   **'undermine the stability of Species':** Barlow, 'Darwin's Ornithological Notes,' 262.

74   **There are four species of mockingbird:** Paquita Hoeck et al., 'Differentiation with Drift: A Spatio-temporal Genetic Analysis of Galápagos Mockingbird Populations (*Mimus* spp.),' *Philosophical Transactions of the Royal Society of London Series B: Biological Sciences* 365 (2010): 1127–1138.

76   **'We see this on every mountain':** Darwin, *On the Origin of Species*, 403.

76   **In his 1947 book *Darwin's Finches*:** David Lack, *Darwin's Finches*, 2nd ed. (Cambridge: Cambridge University Press, 1983). The earliest known description of the Galápagos finches as Darwin's was Percy Lowe, 'The Finches of the Galápagos in Relation to Darwin's Conception of Species,' *Ibis* 78 (1936): 310–321.

76   **In preparation for his visit to the Galápagos:** Lack, *Darwin's Finches*, 1.

76   **The Galápagos finches were not much:** Ibid., 11.

77   **ornithologists following in Lack's footsteps:** Koster and Koster, 'Twelve Days.'

78   **The lineage that led to the warbler finch:** K. Petren et al., 'Comparative Landscape Genetics and the Adaptive Radiation of Darwin's Finches: The Role of Peripheral Isolation,' *Molecular Ecology* 14 (2005): 2943–2957.

79   **'In a sense, we feel we are the bearers':** Peter Grant and Rosemary Grant, *How and Why Species Multiply: The Radiation of Darwin's Finches* (Princeton, NJ: Princeton University Press, 2007), xvii.

79   **'This is one of the most intensive and valuable':** Jonathan Weiner, *The Beak of the Finch: A Story of Evolution in Our Time* (New York: Alfred A. Knopf, 1994), 9.

79   **In the space of just a few months:** Peter Boag and Peter Grant, 'Intense Natural Selection in a Population of Darwin's Finches (*Geospizinae*) in the Galápagos,' *Science* 214 (1981): 82–85; see also Grant and Grant, *How and Why Species Multiply*, 52–54.

79   **As beak size and shape:** Arhat Abzhanov et al., 'The Calmodulin Pathway and Evolution of Elongated Beak Morphology in Darwin's Finches,' *Nature* 442 (2006): 563–567.

79   **today's medium ground finches on Daphne Major:** Grant and Grant, *How and Why Species Multiply*, 57.

81   **A neat experiment conducted in the early 1980s:** Laurene Ratcliffe and Peter Grant, 'Species Recognition in Darwin's Finches (*Geospiza*, *Gould*). II. Geographic Variation in Mate Preference,' *Animal Behaviour* 31 (1983): 1154–1165.

81   **By playing back sound recordings:** Laurene Ratcliffe and Peter Grant, 'Species Recognition in Darwin's Finches (*Geospiza*, *Gould*). III. Male Responses to Playback of Different Song Types, Dialects and Heterospecific Songs,' *Animal Behaviour* 33 (1985): 290–307.

81   **On Daphne Major between 1976 and 1982:** See Grant and Grant, *How and Why Species Multiply*, 97.

82   **'In some, resemblance seems to go for nothing':** Letter from Charles Darwin to Joseph Hooker, 24 December 1856. See Darwin Correspondence Database, http://www.darwinproject.ac.uk/entry-2022 (accessed 31 August 2013).

82   **'When a tortoise is killed':** Gould, 'Birds,' 25.

82   **Recent genetic work shows:** Jennifer Bollmer et al., 'Phylogeography of the Galápagos Hawk (*Buteo galapagoensis*): A Recent Arrival to the Galápagos Islands,' *Molecular Phylogenetics and Evolution* 39 (2006): 237–247.

82   **Darwin quickly realised:** Charles Darwin (1837–1838), Notebook B, CUL-DAR208.5; see darwin-online.org.uk.

84   **This much is obvious from the number of birds:** Darwin may have encountered one of these young or inexperienced hawks on Santiago, counting as many as thirty perched in the scrub around his tent.

84   **on Santiago, where the Galápagos National Park Service:** Christian
      Lavoie et al., *The Thematic Atlas of Project Isabela* (Puerto Ayora: Charles
      Darwin Foundation, 2007).

84   **Without a steady supply of carcasses:** Jose Rivera-Parra et al., 'Implica-
      tions of Goat Eradication on the Survivorship of the Galápagos Hawk,'
      Journal of Wildlife Management 76 (2012): 1197–1204.

85   **with mammal eradication nearing completion:** Josh Donlan et al., 'Recov-
      ery of the Galápagos Rail (*Laterallus spilonotus*) Following the Removal of
      Invasive Mammals,' *Biological Conservation* 138 (2007): 520–524.

85   **The intensity of this colouration:** D. L. Fox, 'Metabolic Fractionation,
      Storage and Display of Carotenoid Pigments by Flamingoes,' *Compara-
      tive Biochemistry and Physiology* 6 (1962): 1–40.

85   **A recent study shows:** Jaime Chaves et al., 'Origin and Population His-
      tory of a Recent Colonizer, the Yellow Warbler in Galápagos and Cocos
      Islands,' *Journal of Evolutionary Biology* 25 (2012): 509–521.

85   **'The place is like a new creation':** Lord George Gordon Byron, *Voyage
      of H.M.S. Blonde to the Sandwich Islands in the Years 1824–1825* (Lon-
      don: John Murray, 1826), 91.

86   **'The birds are Strangers to Man':** Keynes, *Charles Darwin's* Beagle
      Diary, 353, entry for 17 September 1835.

86   **'It would appear that the birds':** Darwin, *Journal of Researches*, 399.

86   **'they would always sidle out of reach':** Beebe, *Galápagos*, 98–99.

87   **'Once we were taught that the earth':** Ibid., 65–66.

Chapter 7

89   **'We must admit that there is no other quarter':** Darwin, *Journal of
      Researches*, 390.

89   **'Little but reptile life is here found':** Melville, 'The Encantadas.' This
      quotation comes from his first sketch, 'The Isles at Large'.

90   **'The black Lava rocks on the beach':** Keynes, *Charles Darwin's* Beagle
      Diary, 353, entry for 17 September 1835.

90   **On Fernandina in 1825:** Byron, *Voyage of H.M.S. Blonde*, 92.

90   **In a rather poetic contribution:** All Thomas Bell's observations on the
      marine iguana come from Thomas Bell, 'On a New Genus of Iguanidae,'
      Zoological Journal 2 (1825): 204–207.

91   **he found it 'largely distended':** Darwin, *Journal of Researches*, 386.

92   **'that when frightened it will not enter the water':** Ibid.

92   **Most marine iguanas do their swimming:** Krisztina Trillmich and Fritz
      Trillmich, 'Foraging Strategies of the Marine Iguana, *Amblyrhynchus cri-
      status*,' *Behavioural Ecology and Sociobiology* 18 (1986): 259–266.

93   **It takes a male marine iguana:** Martin Wikelski and Silke Baurle, 'Pre-copulatory Ejaculation Solves Time Constraints During Copulations in Marine Iguanas,' *Proceedings of the Royal Society of London Series B: Biological Sciences* 263 (1996): 439–444.

93   **Thirty years ago, during the extreme El Niño:** J. E. Cooper and W. A. Laurie, 'Investigation of Deaths in Marine Iguanas (*Amblyrhynchus cristatus*) on Galápagos,' *Journal of Comparative Pathology* 97 (1987): 129–136.

93   **researchers noticed something even more startling:** Martin Wikelski et al., 'Marine Iguanas Shrink to Survive El Niño,' *Nature* 403 (2000): 37–38.

94   **'when, an hour afterwards':** Darwin, *Journal of Researches*, 386.

94   **In the early 1960s, a pair of zoologists:** George Bartholomew and Robert Lasiewski, 'Heating and Cooling Rates, Heart Rate and Simulated Diving in the Galápagos Marine Iguana,' *Comparative Biochemistry and Physiology* 16 (1965): 573–582.

95   **'This animal clearly belongs to the same genus':** Keynes, *Charles Darwin's Zoology Notes*, 336.

95   **In 1983, a couple of immunologists:** J. S. Wyles and Vincent Sarich. 'Are the Galápagos Iguanas Older Than the Galápagos? Molecular Evolution and Colonization Models for the Archipelago,' in *Patterns of Evolution in Galápagos Organisms*, ed. R. I. Bowman et al. (San Francisco: American Association for the Advancement of Science, Pacific Division, 1983), 177–185.

95   **Subsequent studies:** Kornelia Rassmann, 'Evolutionary Age of the Galápagos Iguanas Predates the Age of the Present Galápagos Islands,' *Molecular Phylogenetics and Evolution* 7 (1997): 158–172.

96   **'From their low facial angle':** For quotations regarding the land iguana, see Darwin, *Journal of Researches*, 387–390.

97   **Of the ones Darwin saw:** Keynes, *Charles Darwin's Zoology Notes*, 295.

98   **'When attentively watching an intruder':** Ibid.

98   **strongly suggesting that C. *marthae*:** Gabriele Gentile et al., 'An Overlooked Pink Species of Land Iguana in the Galápagos,' *Proceedings of the National Academy of Sciences* 106 (2009): 507–511.

98   **A more recent expedition, in 2012:** 'DPNG continúa monitoreo poblacional de iguanas rosadas,' press release, Galápagos National Park, 18 September 2012, http://www.galapagospark.org/boletin.php?noticia=678.

98   **But at some point en route to Tahiti:** Sulloway, 'Darwin's Conversion.'

98   **'I have not as yet noticed':** Darwin, *Journal of Researches*, 393.

100   **In a two-hundred-page manuscript:** John Van Denburgh, *The Gigantic Land Tortoises of the Galápagos Archipelago* (San Francisco: Britton & Rey, 1914).

100   **'a species entirely distinct':** David Porter, *Journal of a Cruise Made to the Pacific Ocean* (New York: Wiley & Halstead, 1822), 215.

102   **'Surrounded by the black Lava':** Keynes, *Charles Darwin's* Beagle *Diary*, 354, entry for 21 September 1835

102   **'The tortoises which live on those islands':** Darwin, *Journal of Researches*, 382.

102   **'well beaten roads':** Keynes, *Charles Darwin's* Beagle *Diary*, 362, entry for 9 October 1835.

103   **By munching and trampling:** James Gibbs et al., 'Giant Tortoises as Ecological Engineers: A Long-Term Quasi-experiment in the Galápagos Islands,' *Biotropica* 42 (2010): 208–214.

103   **The obvious way to get a better feel:** Stephen Blake et al., 'Seed Dispersal by Galápagos Tortoises,' *Journal of Biogeography* 39 (2012): 1961–1972.

103   *mezclados*: The *mezclados* are tortoises of unknown provenance and their hybrid descendants that were contained at the Charles Darwin Research Station to prevent them spreading their 'unpure' genes amongst the 'pure' populations in the wild. This explains why they were sterilized prior to release onto Pinta.

104   **The discovery that their closest living relative:** Agaldisa Caccone et al., 'Origin and Evolutionary Relationships of Giant Galápagos Tortoises,' *Proceedings of the National Academy of Sciences* 96 (1999): 13223–13228.

104   **The genetic evidence certainly suggests as much:** Adalgisa Caccone et al., 'Phylogeography and History of Giant Galápagos Tortoises,' *Evolution* 56 (2002): 2052–2066.

105   **The genetic evidence suggests that two independent:** Edgar Benavides et al., 'Island Biogeography of Galápagos Lava Lizards (Tropiduridae: *Microlophus*): Species Diversity and Colonization of the Archipelago,' *Evolution* 63 (2009): 1606–1626.

Chapter 8

108   **'The Lord fill Your Sacred Majesty with holy love':** Letter from Fray Tomás de Berlanga, the bishop of Panama, to His Sacred Imperial Catholic Majesty King Charles I of Spain, 26 April 1535, http://www.galapagos.to/TEXTS/BERLANGA.HTM.

108   **In the Geography and Map Division:** Anonymous (ca. 1565), 'Portolan Chart of the Pacific Coast from Guatemala to Northern Peru with the

Galápagos Islands,' Library of Congress Geography and Map Division, Washington, DC, http://hdl.loc.gov/loc.gmd/g4802c.ct001180.

108   **as historian John Woram has made clear:** John Woram, 'On the Origin of "Galápago,"' http://www.galapagos.to/TEXTS/GALAPAGO.HTM.

109   **'They are extraordinary large and fat':** William Dampier, *A New Voyage Around the World* (London: J. Knapton, 1699), 102.

109   **praises of their 'excellent broth':** Colnett, *A Voyage to the South Atlantic*, 56.

109   **'their flesh, without exception':** Amaso Delano, *A Narrative of Voyages and Travels in the Northern and Southern Hemispheres* (Boston: E. G. House, 1817), 378.

109   **'Hideous and disgusting':** Porter, *Journal*, 151.

109   **Darwin judged tortoise meat to be 'indifferent food':** Keynes, *Charles Darwin's Beagle Diary*, 362, entry for 9 October 1835.

109   **According to Dampier, one party:** Dampier, *A New Voyage*, 109.

109   **'it was common to take out of one of them':** Delano, *A Narrative*, 378.

109   **'does not possess that cloying quality':** Porter, *Journal*, 151.

109   **'They carry with them a constant supply':** Ibid.

110   **'They were piled up on the quarter-deck':** Ibid., 214.

110   **'fresh provisions for six or eight months':** Morrell, *A Narrative of Four Voyages*, 125.

110   **'All hands employed in making belts':** See Charles Haskins Townsend, 'The Galápagos Tortoises in Their Relation to the Whaling Industry,' *Zoologica* 4 (1925): 55–135.

110   **'Four boats were dispatched':** Porter, *Journal*, 214.

110   **On another occasion he recorded:** Porter, *Journal*, 162.

110   **In an analysis way ahead:** Townsend, 'The Galápagos Tortoises.' Townsend's comprehensive survey contains the quotations from the sailors.

112   **In around 1820:** Graham Burnett, *Trying Leviathan: The Nineteenth-Century New York Court Case That Put the Whale on Trial and Challenged the Order of Nature* (Princeton, NJ: Princeton University Press, 2010).

113   **'The appearance of this man':** Porter, *Journal*, 131.

113   **'He struck strangers much as if':** Melville, 'The Encantadas,' published in *Putnam's Monthly Magazine*. This quotation comes from his ninth sketch, 'Hood's Isle and the Hermit Oberlus'.

113   **'We have seen, from what Patrick':** Porter, *Journal*, 232.

114   **'the productions of the island':** Jeremiah Reynolds, *Voyage of the United States Frigate Potomac: Under the Command of Commodore John Downes* (New York: Harper and Brothers, 1835), 467.

114   **'Surrounded by tropical vegetation':** FitzRoy, *Narrative*, 491.

115   **Reynolds also noted:** Reynolds, *Voyage*, 467.

115   **'what havoc the introduction of any new beast':** Darwin, *Journal of Researches*, 401.

115   **there were also 536 introduced invertebrates:** Rachel Atkinson et al., 'Fifty Years of Eradication as a Conservation Tool in Galápagos,' in *The Role of Science for Conservation*, ed. Matthias Wolf and Mark Gardener (London: Routledge, 2012), 183–198.

115   **870 species, according to the latest reckoning:** Ibid.

115   **A recent study of satellite images:** James Watson et al., 'Mapping Terrestrial Anthropogenic Degradation on the Inhabited Islands of the Galápagos Archipelago,' *Oryx* 44 (2010): 79–82.

116   **Their total haul of more than 75,000 specimens:** See Edward Larson, *Evolution's Workshop* (London: Penguin, 2001), 119–144.

116   **If it appears as though these men:** See Matthew James, 'The Boat, the Bay, and the Museum,' in *The Role of Science for Conservation*, ed. Matthias Wolf and Mark Gardener (London: Routledge, 2012), 87–99.

117   **'to make an exhaustive survey':** Joseph Slevin, 'Log of the Schooner "Academy,"' *Occasional Papers of the California Academy of Sciences* 17 (1931): 5.

Chapter 9

119   **'The Galápagos did not make Darwin':** Frank Sulloway, 'Darwin and the Galápagos,' *Biological Journal of the Linnean Society* 21 (1984): 29–59.

119   **In comparison to Darwin-based celebrations:** For details on the 1909 celebrations, see John van Wyhe, 'Where Do Darwin's Finches Come From? *Evolutionary Review* 3 (2012): 185–195.

120   **For historian Edward J. Larson:** Larson, *Evolution's Workshop*, 152.

120   **'general grasp and sheer interest':** Beebe, *Galápagos*, 428.

120   **Most infamous amongst these new arrivals:** For the best account of the bizarre goings-on on Floreana in the 1930s, read John Treherne, *The Galápagos Affair* (London: Jonathan Cape, 1983).

120   **'Raising the monument was more':** Victor von Hagen, *Ecuador the Unknown & the Galápagos Revisited* (Oxford: Oxford University Press, 1940). Von Hagen's Darwin monument is now within the grounds of the naval base on San Cristóbal.

120   **Ecuador had already made moves:** Ecuador's Registro Oficial No. 257 of 31 August 1934.

121   **On 4 July that year:** The creation of the Galápagos National Park occurred several years before other, supposedly highly developed

nations got around to protecting their own natural treasures. France, for instance, only did so in 1963, Germany in 1970, and Belgium in 2006.

121 **'to provide knowledge and assistance':** For the Charles Darwin Foundation's mission and vision, see http://www.darwinfoundation.org/about-us/mission-and-vision.

121 **'the preservation of all sources of pure wonder':** Julian Huxley, *The Humanist Frame* (New York: Harper & Brothers, 1961), 52.

121 **an annual operating budget of around $15 million:** Magaly Oviedo, presentation in London, 2011. In 2010, GNPS spent $14,112,340.

123 **Some of these hybrids show:** Caccone et al., 'Phylogeography and History.'

123 **It also turns out that there are descendants:** Nikos Poulakakis et al., 'Historical DNA Analysis Reveals Living Descendants of an Extinct Species of Galápagos Tortoise,' *Proceedings of the National Academy of Sciences* 105 (2008): 15464–15469.

123 **Several tortoises even have a smattering:** Danielle Edwards et al., 'The Genetic Legacy of Lonesome George Survives: Giant Tortoises with Pinta Island Ancestry Identified in Galápagos,' *Biological Conservation* 157 (2013): 225–228.

123 **Project Isabela—an $8.5-million initiative:** Lavoie et al., *The Thematic Atlas of Project Isabela.*

124 **Trials revealed that the most effective:** Karl Campbell et al., 'Eradication of Feral Goats *Capra hircus* from Pinta Island, Galápagos, Ecuador,' *Oryx* 38 (2004): 328–333; Karl Campbell et al., 'Increasing the Efficacy of Judas Goats by Sterilisation and Pregnancy Termination,' *Wildlife Research* 32 (2005): 737–743.

124 **sterilised females treated with a cocktail:** Karl Campbell et al., 'Development of a Prolonged Estrus Effect for Use in Judas Goats,' *Science* 102 (2007): 12–23.

124 **somewhere in the region of 200,000:** Victor Carrion et al., 'Archipelago-wide Island Restoration in the Galápagos Islands: Reducing Costs of Invasive Mammal Eradication Programs and Reinvasion Risk,' *PLoS One* 6 (2011): e18835.

124 **The GNPS put the cost of removing:** Ibid.

125 **Project Pinzón:** 'Proyecto Pinzón: Restauración de los ecosistemas de las Islas Galápagos mediante la eliminación de roedores introducidos,' Reporte Final del Taller de Galápagos: Desarrollando una Estrategia para la Eradicación de Roedores Introducidos en el Archipiélago de Galápagos, 2–12 Abril, 2007.

126 **in January 2011, Project Pinzón entered:** See Henry Nicholls, 'The 18-km^2 Rat Trap,' *Nature* 497 (2013): 306–308.

128   **In 2000, the World Conservation Union:** S. Lowe et al., *100 of the World's Worst Invasive Alien Species: A Selection from the Global Invasive Species Database* (World Conservation Union, 2000).

128   **In the Galápagos, it has been observed:** Y. D. Lubin, 'Changes in the Native Fauna of the Galápagos Islands Following Invasion by the Little Red Fire Ant, *Wasmannia auropunctata*,' *Biological Journal of the Linnean Society* 21 (1984): 229–242.

128   **'On the larger islands':** Charlotte Causton et al., 'Eradication of the Little Fire Ant, *Wasmannia auropunctata* (Hymenoptera: Formicidae), from Marchena Island, Galápagos: On the Edge of Success?' *Florida Entomologist* 88 (2005): 159–168.

129   **When a new infestation turned up in 2008:** Atkinson et al., 'Fifty Years of Eradication."

129   **On Santa Cruz, the quinine tree:** Christopher Buddenhagen et al., 'The Control of a Highly Invasive Tree *Cinchona pubescens* in Galápagos,' *Weed Technology* 18 (2004): 1194–1202.

129   **the invasion of this one plant:** Heinke Jager et al., 'Tree Invasion in Naturally Treeless Environments: Impacts of Quinine (*Cinchona pubescens*) Trees on Native Vegetation in Galápagos,' *Biological Conservation* 140 (2007): 297–307; Heinke Jager et al., 'Destruction Without Extinction: Long-Term Impacts of an Invasive Tree Species on Galápagos Highland Vegetation,' *Journal of Ecology* 97 (2009): 1252–1263.

130   **But it's expensive:** Christopher Buddenhagen et al., 'The Cost of Quinine *Cinchona pubescens* Control on Santa Cruz Island, Galápagos,' *Galápagos Research* 63 (2005): 32–36.

130   **Of thirty such projects:** Mark Gardener et al., 'Eradications and People: Lessons from the Plant Eradication Program in Galápagos,' *Restoration Ecology* 18 (2010): 20–29.

130   **the presence of little fire ants:** Charlotte Causton, 'Dossier on *Rodolia cardinalis* Mulsant (Coccinellidae: Cocinellinae), a Potential Biological Control Agent for the Cottony Cushion Scale, *Icerya purchasi* Maskell (Margarodidae) (Puerto Ayora: Charles Darwin Foundation, 2001).

131   **A decade on:** Carolina Calderón Alvarez, 'Monitoring the Effects of *Rodolia cardinalis* on *Icerya purchasi* Populations on the Galápagos Islands,' *BioControl* 57 (2012): 167–179; Mark Hoddle et al., 'Post Release Evaluation of *Rodolia cardinalis* (Coleoptera: Coccinellidae) for Control of *Icerya purchasi* (Hemiptera: Monophlebidae) in the Galápagos Islands,' *Biological Control* 67 (2013): 262–274.

131   **If there were any doubt:** Birgit Fessl and Sabine Tebbich, *Philornis downsi*—a Recently Discovered Parasite on the Galápagos Archipelago—a Threat for Darwin's Finches? *Ibis* 144 (2002): 445–451.

131   **It is pretty well established:** David Wiedenfeld et al., 'Distribution of the Introduced Parasitic Fly *Philornis downsi* (Diptera, Muscidae) in the Galápagos Islands,' *Pacific Conservation Biology* 13 (2007): 14–19.

131   **Several studies reveal the devastating:** Rachel Dudaniec and Sonia Kleindorfer, 'Effects of the Parasitic Flies of the Genus *Philornis* (Diptera: Muscidae) on Birds,' *Emu* 106 (2006): 13–20; Rachel Dudaniec et al., 'Effects of the Introduced Ectoparasite *Philornis downsi* on Haemoglobin Level and Nestling Survival in Darwin's Small Ground Finch (*Geospiza fuliginosa*),' *Austral Ecology* 31 (2006): 88–94.

132   **Several cargo ships zip back and forth:** For a thorough look at quarantine in the Galápagos, see WILDAID, 'The Quarantine Chain: Establishing an Effective Biosecurity System to Prevent the Introduction of Invasive Species into the Galápagos Islands,' WILDAID, 2012, http://is.gd/lbEOfJ.

132   **'The Americans purchased all that could be fished':** Stein Hoff, *The Galápagos Dream: An Unknown History of Norwegian Emigration*, trans. Mrs. Friedel Horneman (Oslo: Grødahl & Søn Forlag A.s., 1985). See http://www.galapagos.to/TEXTS/HOFF-1.PHP.

132   **the airstrip made it possible:** For an extremely clear analysis of the opening of the Galápagos, see Christophe Grenier, 'Nature and the World: A Geohistory of Galápagos,' in *The Role of Science for Conservation*, ed. Matthias Wolf and Mark Gardener (London: Routledge, 2012), 256–274.

132   **Beyond the little white posts:** See Michael Romero and Martin Wikelski, 'Exposure to Tourism Reduces Stress-Induced Corticosterone Levels in Galápagos Marine Iguanas,' *Biological Conservation* 108 (2002): 371–374; Luis Fernando De León et al., 'Exploring Possible Human Influences on the Evolution of Darwin's Finches,' *Evolution* 65 (2011): 2258–2272.

Chapter 10

136   **This kind of backward development:** Byron Villacis and Daniela Carrillo, 'The Socioeconomic Paradox of Galápagos,' in *The Role of Science for Conservation*, ed. Matthias Wolf and Mark Gardener (London: Routledge, 2012), 69–85.

136   **The special law is a long document:** A PDF of the Special Law for Galápagos is available on the Galápagos National Park website at http://is.gd/6uAGPW.

137   **'At the GNPS, the situation was worse still':** A copy of the UNESCO report *Joint IUCN/UNESCO Mission Report: Galápagos Islands (Ecuador)* (presented at the thirtieth session of the World Heritage Committee in

Vilnius, Lithuania, 8–16 July 2006) can be obtained from the World Heritage Centre at http://is.gd/nEsQhI.

137 **'the bureaucratization of the Galápagos':** See Epler, *Tourism*, and Graham Watkins and Alejandro Martinez, *Galápagos Report 2007–2008: The Changing Organizational Framework in Galápagos* (Puerto Ayora: Charles Darwin Foundation, 2008).

138 **All of this was of considerable concern:** UNESCO, *Joint IUCN/UNESCO Mission Report*.

138 **The Inspection and Quarantine System:** WILDAID, 'The Quarantine Chain.'

138 **There was a rapid increase in the number:** Epler, *Tourism*.

138 **Conclusion: 'Galápagos is shifting':** UNESCO, *Joint IUCN/UNESCO Mission Report*.

139 **'On the contrary,' they reported:** UNESCO, *Report of the Reactive Monitoring Mission: Galápagos Islands (Ecuador)* (presented at the thirty-first session of the World Heritage Committee in Christchurch, New Zealand, 23 June–2 July 2007).

139 **On 10 April, in the middle:** Ibid.

139 **a measure intended 'to draw attention':** Ibid.

139 **In the opinion of leading conservationists:** Graham Watkins and Felipe Cruz, *Galápagos at Risk: A Socioeconomic Analysis* (Puerto Ayora: Charles Darwin Foundation, 2007).

139 **In recognition of these steps:** UNESCO, *Report of the Decisions Adopted by the Thirty-Fourth Session of the World Heritage Committee, Brasilia, Brazil, 25 July–3 August 2010* (Paris: World Heritage Committee, 2010).

139 **It's been estimated that around 10 percent:** Epler, *Tourism*.

140 **Between 1999 and 2005:** Edward Taylor et al., *Ecotourism and Economic Growth in the Galápagos: An Island Economy-wide Analysis* (Davis: University of California, Davis, 2006).

140 **the employed majority is able to earn:** Villacis and Carrillo, 'The Socioeconomic Paradox.' In 2009, there was 4.9 percent unemployment in the Galápagos compared to 7.9 percent in mainland Ecuador. Average income across both public and private sectors stood at $772.03 per month in the Galápagos against just $251.70 on the continent.

140 **With few permanent residents interested:** Walter Jimbo and Christophe Grenier, 'The Construction Sector of Puerto Ayora,' in *Galápagos Report 2009–2010* (Puerto Ayora: GNPS, GCREG, CDF, and GC, 2010).

140 **In 2007, there were thirty-eight marriages:** Villacis and Carrillo, 'The Socioeconomic Paradox,' 72.

141   **a 2009 survey of over 1,000 households:** Laura Brewington, 'The Double Bind of Tourism in Galápagos Society,' in *Science and Conservation in the Galápagos Islands: Frameworks and Perspectives*, ed. Stephen Walsh and Carlos Mena (New York: Springer, 2012), 105–125.

141   **rapid immigration has radically altered:** Villacis and Carrillo, 'The Socioeconomic Paradox,' 74, Figure 4.4.

141   **Into this painful social transition walked:** The description of Raquel Molina's directorship of the Galápagos National Park Service is based on news stories and Douglas Griffiths, 'Galápagos National Park Director Firing May Threaten Conservation Efforts,' confidential memo sent from Guayaquil to Washington, DC, 28 March 2008, 09GUAYA-QUIL77. See http://is.gd/V9rtMO.

144   **'No area on Earth of comparable size':** Bowman, 'Contributions to Science from the Galápagos,' 278.

Appendix A

147   **'There are lots of people who come to Galápagos':** Magaly Oviedo, personal communication, 6 July 2011.

147   **If this lengthy amendment:** Letter from the Standing Specialised Committee of the Autonomous Governments, Decentralization, Competencies and Territorial Arrangement to Fernando Cordero Cueva, President of the National Assembly, Republic of Ecuador, 13 June 2012.

147   **The fee structure would also begin operating:** These changes to the fee structure were outlined by Magaly Oviedo at a workshop in London on 6 July 2011. See Abigail Rowley, 'Ecuador and Galápagos: Ensuring a Successful Future for Tourism, Local People and Wildlife,' Galápagos Conservation Trust, July 6, 2011, http://www.savegalapagos.org/galapagos/Report%20FINAL%20-%20Nov%202011.pdf.

149   **In 2011, there were ninety-three vessels:** Parque Nacional Galapagos, *Listado de Patentes Vigentes Para el Período 2011–2–1 hasta 2012–1–31.* See http://is.gd/cAjXEl.

150   **In 1982, there were just eighteen hotels:** Epler, *Tourism.*

Further Reading

Of the general guidebooks to the natural history of the islands, there are several that stand out: *Galápagos: A Natural History* by Michael Jackson (University of Calgary Press, 1998); *Wildlife of the Galápagos* by Julian Fitter, Daniel Fitter and David Hosking (Collins, 2007); and *Galápagos Wildlife* by David Horwell and Pete Oxford (Bradt Travel Guides, 2011).

For the scientific importance of the Galápagos, see *Evolution's Workshop: God and Science in the Galápagos Islands* by Edward Larson (Penguin Books Ltd, 2002). William Beebe's *Galápagos: World's End* (Putnam's Sons, 1924) is still a wonderful read almost a century on. In *The Arcturus Adventure* (Putnam's Sons, 1926), Beebe spends more time in the water, making this a must-read for anyone with an interest in diving. The importance of Darwin's finches is captured quite brilliantly in Jonathan Weiner's Pulitzer Prize-winning *The Beak of the Finch: A Story of Evolution in Our Time* (Knopf, 1994). For a more recent and more scientific take on these iconic birds, get a copy of the Peter and Rosemary Grant's *How and Why Species Multiply* (Princeton University Press, 2011). *A Sheltered Life: The Unexpected History of the Giant Tortoise* by Paul Chambers and *Lonesome George: The Life and Loves of a Conservation Icon* by me (Macmillan Science, 2006) both focus on the tortoises that gave the islands their name. For children, *Island: A Story of Galápagos* by Jason Chin (Roaring Book Press, 2012) is a perfect introduction to the processes that led to the formation of the Galápagos.

There has been much work on what Charles Darwin thought about the Galápagos, notably a series of cleverly argued articles written by Frank Sulloway in the 1980s and Janet Browne's wonderful biography *Charles Darwin: Voyaging* (Knopf, 1995). More recently, Thalia Grant and Gregory Estes have published *Darwin in Galapagos: Footsteps to a New World* (Princeton University Press, 2009), the definitive work on the subject. All Darwin's published and unpublished works can be found online at *Darwin Online* (darwin-online.org.uk), his letters at the *Darwin Correspondence Project* (www.darwinproject.ac.uk) and other key historical texts at *Las Encantadas: Human and Cartographic History of the Galápagos Islands* (www.galapagos.to).

There are several excellent treatments of the human side of the Galápagos, especially *The Enchanted Islands* by John Hickman (Anthony Nelson, 1991) and *Charles Darwin Slept Here* by John Woram (Rockville Press, 2005). *The Galapagos Affair* by John Treherne is the gripping real-life mystery that unfolded on Floreana in the 1930s. For an exploration of more recent social change, see *Plundering Paradise: The Hand of Man on the Galápagos Islands* by Michael D'Orso (HarperCollins, 2002) or *Galapagos at the Crossroads: Pirates, Biologists, Tourists and Creationists Battle for Darwin's Cradle of Evolution* by Carol Ann Basset (National Geographic Society, 2009).

In the coffee-table genre, I have selected just three: *Galápagos: The Islands That Changed the World* by Paul Stewart and Richard Dawkins (BBC Books, 2006), accompanied the BBC's three-part documentary on the islands; *Galapagos: Preserving Darwin's Legacy* (Firefly Books, 2009), an edited volume by photographer Tui De Roy with chapters written by leading scientists and conservationists; and *Galapagos: Both Sides of the Coin* by Pete Oxford and Graham Watkins (Imagine Publishing, 2009), which depicts both the natural wonders and the human stresses on the archipelago.

Index